Microsoft
Office
Word 2007
A Beginners Guide

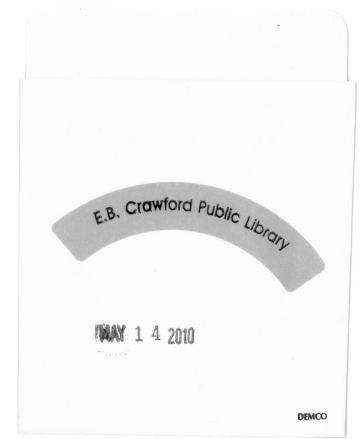

By WR Mills

AuthorHouse™
1663 Liberty Drive
Bloomington, IN 47403
www.authorhouse.com
Phone: 1-800-839-8640

First published by AuthorHouse 1/13/2010

IBSN: 978-1-4490-3238-8 (e)
ISBN: 978-1-4490-3237-1 (sc)

Library of Congress Control Number: 2010900287

Printed in the United States of America
Bloomington, Indiana

This book is printed on acid-free paper.

Microsoft

Office

Word 2007

A Beginners Guide

A training book for Microsoft Word 2007

By WR Mills

About the Author

Bill has a background in electronics and technology. He started writing software in 1982 and has expanded his programming skills to include C, C++, and Visual Basic. Bill also designs web sites. He designed a computer based telephone system for the hotel/motel market.

In 2007 he started teaching computer training classes and seems to have a knack for explaining things in a simple way that the average user can understand.

Bill is self-employed and lives in Branson Missouri with his wife Rose. They have three children, two sons and a daughter.

Preface

In 2007 I started teaching computer training classes. I was shocked at how much trouble the students had trying to understand the textbooks. I spent all of my time explaining what the textbook was trying to get across to the reader. It wasn't until I started getting ready for teaching the Microsoft Office 2007 series of classes that I finally gave up and started writing the textbooks myself.

These books are easy to understand and have step by step, easy to follow, directions. These books are not designed for the computer geek; they are designed for the normal everyday user.

It seems I have a knack for explaining things in a simple way that the average user can understand. I hope this book will be of help to you.

William R. Mills

Foreword

Dear Bill, I wanted to write a note of appreciation to you for your books: Microsoft Office Word 2007, Microsoft Office Excel 2007, and Microsoft Office PowerPoint 2007. I've used them all and found each one to be easy to read and very user friendly. If anyone needs to learn one of the 3 programs, but is even a little intimated, I strongly suggest they try one of your books. It's almost as good as taking a class with you as the instructor. If I didn't understand a step, I just went back to the pervious step and tried it again --- and it always worked!! There's just enough humor in the text to keep the reading interesting; never dull, but fun and light. Just what a beginner needs. Again, Bill, I thank you for creating these books that make learning something I needed to learn fun and easy. Sincerely, Cyndy O

Important Notice:

There will be times during this book that you be asked to open a specific file for the lesson. These files can be downloaded from the EZ 2 Understand Computer Books web site.

Open your internet browser (probably Internet Explorer) and go to www.ez2understandcomputerbooks.com. Click your mouse on the <u>Lesson Files</u> link toward the top. This will take you to the page where you can download the files needed. There are directions on the page to help you with the download. They are repeated below for your convenience.

To download the files follow the following steps:

1) Right-click your mouse on the file(s) you want. These are zipped files and contain the lessons that you will need for each book.

2) Select the "Save Target As" choice. Make sure the download is pointed to a place on your hard drive where you can find it, such as My Documents.

3) Click the Save button

4) The files are zipped files and will need to be extracted to access the contents. To extract the files right-click on the file and choose "Extract All".

Table of Contents

Chapter One   The Basics

Microsoft Word 2007 is a powerful word processor program. Word also has a variety of tools for the user. These tools will allow the user to produce professional documents. With all of these tools available to you, Word 2007 is extremely easy to use. If you are accustomed to using a previous version of Word you will, after a short time, fall in love with Word 2007.

The first thing that you are going to notice is that Word 2007 looks different than any other version of Word. This is because of the new user interface. You might ask why this version of Word is better than the version you are use to. Let me answer that in this way; do you remember searching through a series of menus and submenus to find a command? That is all a thing of the past. Word 2007 has the Ribbon. Wow! Are you excited yet?

Is the Ribbon scary? Probably. Is it intimidating? More than likely. Is it better and easier to use? Yes definitely. The Ribbon is based more on how people actually use their computer, and not on the way a programmer wanted it to be used.

The Ribbon is divided into Task Orientated Tabs. Each tab has groups of related commands. Everything you need is right at your fingertips. You will not have to search through menus and submenus until you want to pull your hair out, trying to find a command.

Word can automatically check your spelling and grammar, and it can correct common mistakes. For example if you type teh Word will automatically change it to the. With Word you can insert charts, tables, and pictures into your documents.

There are also new "Quick Styles". These are ready-made styles that give your documents a professional makeover fast.

This chapter is an introduction to the basics of Microsoft Word. We will cover what you need to create, print, and save a document. We will also start covering the new Ribbon.

Lesson 1 - 1   Starting Word

Starting Word is a simple process. The first thing you have to do is have your computer on and the desktop showing. I know, I didn't have to say that, but I did, too late to take it back now. The Word program is located in the Microsoft Office folder.

Click on the Start Button

Select All Programs

Move the mouse to Microsoft Office and click on Microsoft Word 2007 from the menu that slides out to the right side

In a few moments, Word will appear on your screen. It should look similar to Figure 1-1. You may not have all of the options at the top; it depends on the available screen width of your monitor.

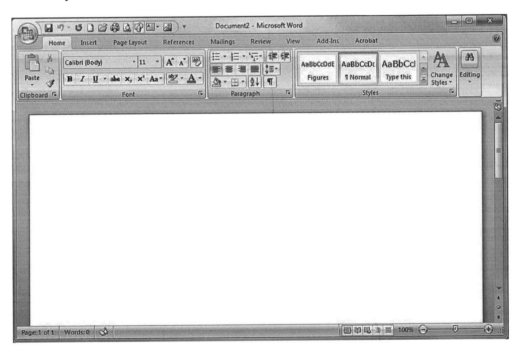

Figure 1-1

Word is now ready for you to start typing your information on the screen. From here you can type information, format the text, and insert objects, such as pictures and charts.

Before we jump headlong into all of the different things that you can do, let's get use to the screen.

Lesson 1 – 2   Understanding the Word Screen

The moment you start Word 2007 you will notice some major changes. Microsoft completely redesigned the user interface. Microsoft pretty much went back to the drawing board to design the way you use Word. How it works now is based on how most people actually use the program. Figure 1-2 shows the Word screen.

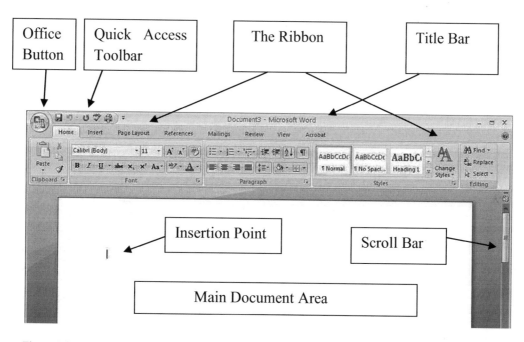

Figure 1-2

Figure 1-3 shows a close-up view of the Office Button. The Office button is actually located on the end of the Quick Access Toolbar. We will discuss the Office button more in the next lesson.

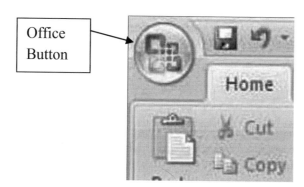

Figure 1-3

Figure 1-4 shows the Quick Access Toolbar. We will discuss this in Lesson 1-5.

Figure 1-4

The Title Bar (at the very top) looks a little different in this version of Word. You will also notice that the Standard and Formatting Toolbars are gone. They have been replaced by the Ribbon and the Quick Access Toolbar. Both the Ribbon and the Quick Access Toolbar are new and will be discussed throughout the remainder of the book.

The Insertion point is where the flashing vertical line is located. This lets the user know where the next character typed will appear in the document.

The Main Document Area is the part of the screen where the text that you type will appear.

The Scroll Bar will allow you to move up and down in the document. You can move the slider up and down as necessary to view other text that is not visible. Some of the text may be on the same page or it can be on a different page in the document.

At the bottom of the screen is the Status Bar, which can display information to the user.

Lesson 1 – 3   The Office Button

The Office Button, for the most part, has taken the place of the old File section of the menu bar. As you can see the menu bar does not exist in this version of Word. In this lesson we will examine the Office Button and see just how it works.

Using your mouse, click on the Office Button

When you click on the Office Button, a menu will drop down giving you several choices of what you are able to do. This is shown in Figure 1-5.

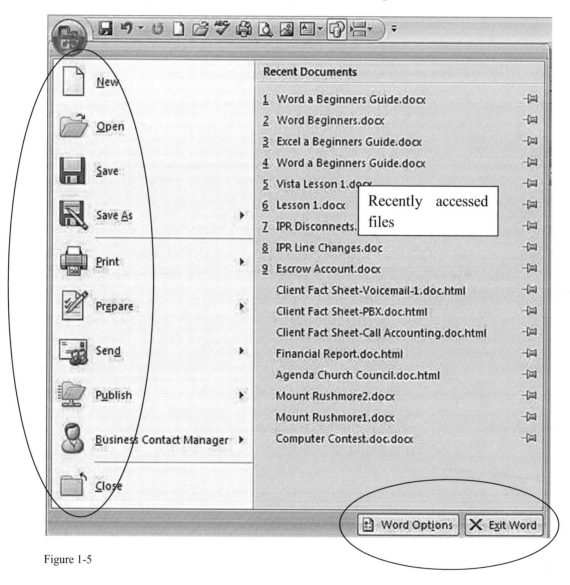

Figure 1-5

On the left side you will notice that many of the choices were the same as when you clicked on the file button of the older style menu bar. You can start a new document or open an existing document. You also have the Save and Save As choices. You may notice that the drop-down choices are divided into two sections. The most popular choices are put at the top. The lesser used options are placed toward the bottom.

If you needed to print the document, you would find the printer options under the print choice.

The Prepare section is where you would look at the properties of the document as well as encrypt it so no one could open it or edit it without a password.

The Send option is where you could send the document as an e-mail or a fax.

The Publish section is where you would share the document with others or create a new site for the document.

In the Business Contact Manager, you can link this document to the communications history of a business record.

On the right side are several of the most recent documents that you have opened. To open one of these documents you simply click on it with the mouse.

The last choice on the left is where you click to close a document.

On the bottom right you will see that you can also exit Word from this part of the menu.

Also on the bottom right is a button to access the Word Options. The first screen of the available options is shown starting in Figure 1-6.

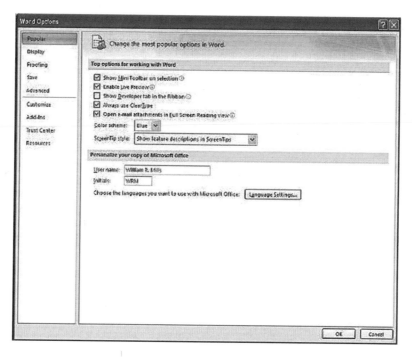

Figure 1-6

From this screen you can change the most popular options, such as should the Mini Toolbar be shown when you select text, and what user name to attach to a document.

In the Display section, shown in Figure 1-7, you can choose such options as whether to show document tooltips and should drawings be printed.

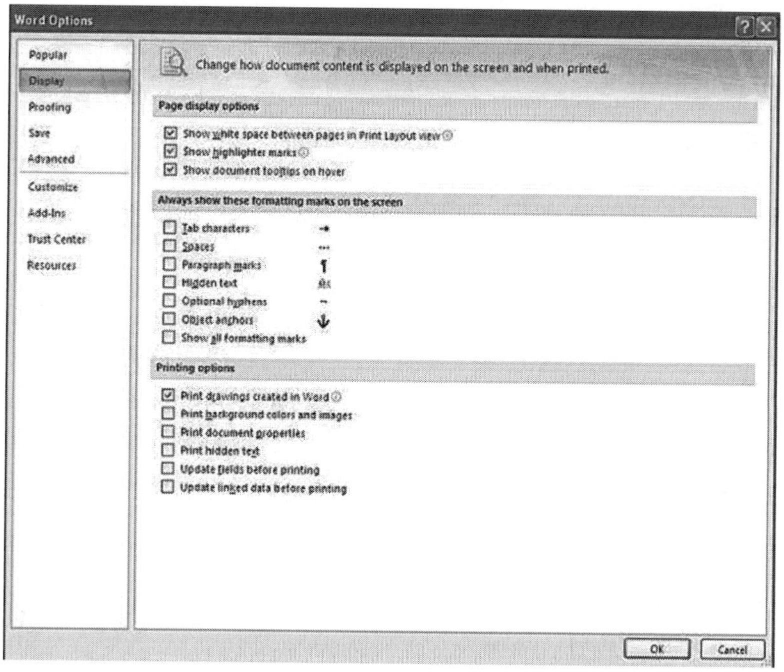

Figure 1-7

7

In the Proofing section, shown in Figure 1-8, you can make changes as to how Word corrects and formats words as you type, and if Word will ignore certain words. You can also tell Word if it should check your grammar when it checks your spelling.

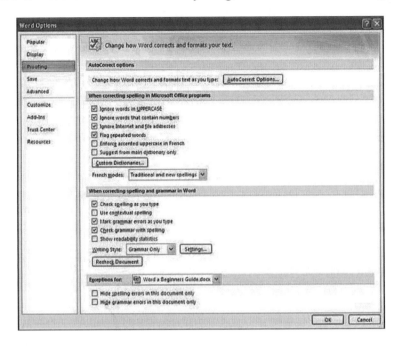

Figure 1-8

In the Save section, you can customize how documents are saved. This screen is shown in Figure 1-9.

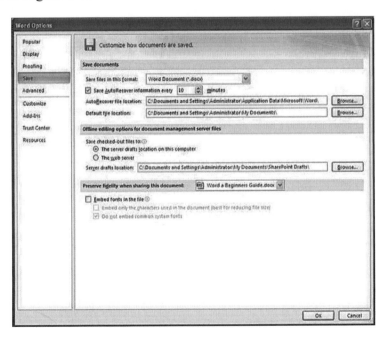

Figure 1-9

In the advanced section you can make changes to some of the editing options, pasting options, other display options, some printing options, and background saves. This screen is shown in Figure 1-10. You can use the scroll bar to see the other advanced options.

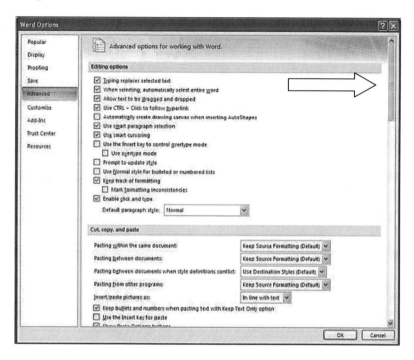

Figure 1-10

The Customize section allows you to add and remove icons from the Quick Access Toolbar. You can also move the toolbar to below the Ribbon instead of having it above the Ribbon. See Figure 1-11 for the Customize screen.

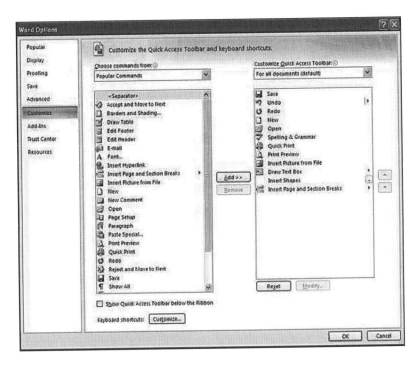

Figure 1-11

The add-ins section shows the "extra" things that have been added to help Word work better. This is shown in Figure 1-12.

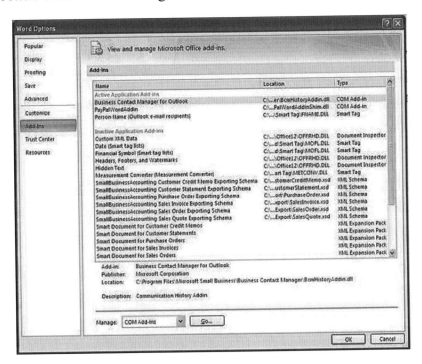

Figure 1-12

10

The Trust Center contains security and privacy settings. Microsoft recommends that you do not change these settings. See Figure 1-13 for the Trust Center.

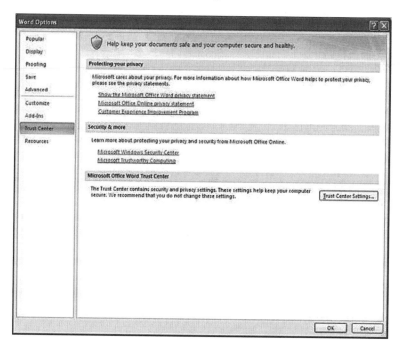

Figure 1-13

The last section of the Word Options is the Resource Center. From here you can get updates, contact Microsoft, etc. this screen is shown in Figure 1-14.

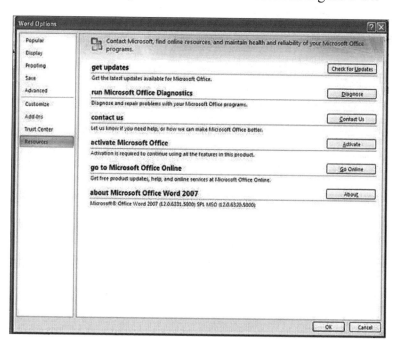

Figure 1-14

Lesson 1 – 4   The Ribbon – An Overview

The Ribbon has been designed to offer an easy access to the commands that you (the user) use most often. You no longer have to search for a command embedded in a series of menus and submenus. The Ribbon has a series of Tabs and each tab is divided into several groups of related commands. Figure 1-15 shows the Ribbon across the top of the Word program.

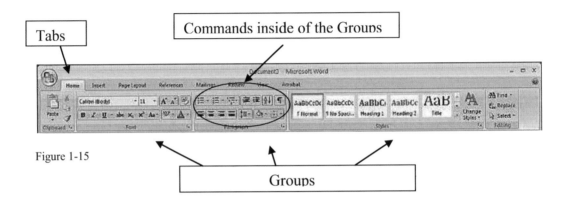

Figure 1-15

There are three major components to the Ribbon.

Tabs:

There are seven basic tabs across the top.

The Home Tab contains the commands that you use most often.

The Insert Tab contains all of the objects that can be inserted into a document.

The Page Layout Tab contains the choices for how each page will look.

The References Tab contains things such as Table of Contents and Footnotes.

The Mailings Tab contains such things as Starting a Mail Merge and creating envelopes and labels.

The Review Tab has things that are related to proofing, protecting, and comments.

The View Tab allows you to change to the different views that are available.

Groups:

Each Tab has several Groups that show related items together.

Look at the Home Tab to see an example of the related Groups.

The Home Tab has the following Groups: Clipboard, Font, Paragraph, Styles, and Editing.

Commands:

A Command is a button, a box to enter information, or a menu.

The Clipboard Group, for example, has the following commands in it: Cut, Copy, Paste, and Format Painter.

Note: You may have an additional tab on the Ribbon. This tab is the Acrobat Tab. If it is available, this is an easy access to the commands that will allow you to save files in the .pdf format.

When you first glance at a group, you may not see a command that was available from the menus of the previous versions of Word. If this is the case you need not worry. Some Groups have a small box with an arrow in the lower right side of the Group. See figure 1-16 for a view of a group with this arrow.

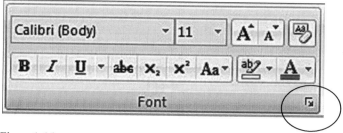

Figure 1-16

This small arrow is called the Dialog Box Launcher. If you click on it, you will see more options related to that Group. These options will usually appear in the form of a Dialog Box. You will probably recognize the dialog box from previous versions of Word. These options may also appear in the form of a task pane. Figure 1-17 shows the Font Dialog Box.

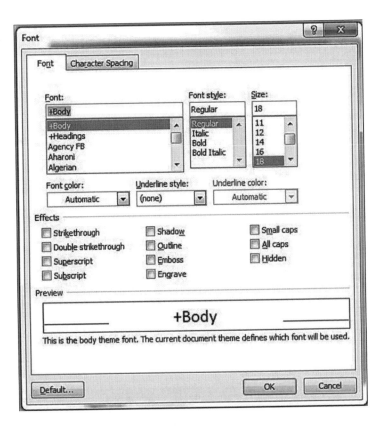

Figure 1-17

Speaking of previous versions, if you are wondering whether you can get the look and feel of the older versions of Word back, the answer is simple, **no you can't**.

The good news is that after playing with and using the Ribbon, you will probably like it even better. It really does make working with the word processor easier. The Ribbon will be used extensively and the tabs will be covered in more detail later as we go through this book.

Lesson 1 – 5   The Quick Access Toolbar

The Ribbon, as you will find out, is wonderful, but what if you want some commands to always be right at your fingertips without having to go from one tab to another? Microsoft gave us a toolbar for just that purpose. This toolbar is called the Quick Access Toolbar and is located just above, or below, and to the left end of the Ribbon. The Quick Access Toolbar is shown in Figure 1-18.

Quick Access Toolbar

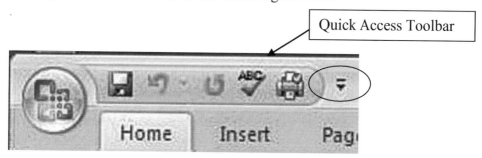

Figure 1-18

The Quick Access Toolbar contains such things as the Save button, the Undo and Redo button, the Quick Print button, and Spell Check button. These are things that you normally use over and over and you will want them available all of the time.

There is even more good news, if you want to add an item to the toolbar, the process is very simple. At the right end of the toolbar is an arrow pointing downwards. If you click on this arrow, a new drop down menu will come onto the screen, as shown in Figure 1-19.

Click on the down arrow on the right side of the Quick Access Toolbar

From this menu you can choose from the standard choices or you can customize the toolbar to suit your needs by clicking on the More Commands choice.

You can also choose to show the tool bar below the Ribbon instead of above it. I have my computer set to show the Quick Access Toolbar below the Ribbon, probably because the toolbars were always below the menu bar in the older versions.

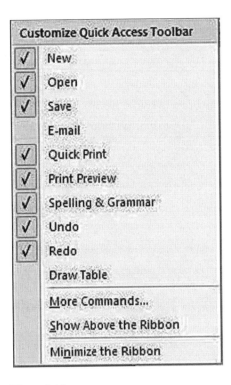

Figure 1-19

If you want to add an item from the standard choices all you have to do is click on the item you want to add. The drop down menu will disappear and the new item will be added to the toolbar.

Figure 1-19 shows most of the items that you will use with all Word documents.

Add the checked items to your Quick Access Toolbar

If the option you want to add is not listed in the standard choices, all of the available options are listed under the More Commands.

Figure 1-20 shows the Word Options Dialog box that will come to the screen if you choose the More Commands option.

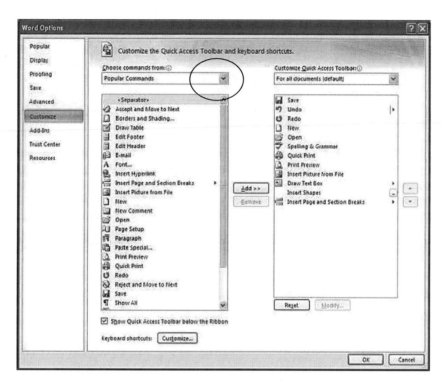

Figure 1-20

If you wish to add an item to the Quick Access Toolbar, all you need to do is click on the option on the left and then click the Add button in the center. When you are finished adding items, click the OK button to place them in the toolbar. You will probably find that there are several things that you will use over and over with every document and you will want to place them in the Quick Access Toolbar just because this will save you so much time.

By default the most popular commands are shown when you open the More Commands dialog box. Clicking the down arrow (where the circle is in the figure) will allow you to choose from the other choices available.

Lesson 1 – 6   Using the Keyboard

What about all of you people who prefer to use the Keyboard over the mouse? I have not forgotten about you, and this lesson is just for you. People who prefer the keyboard over the mouse often started way back with DOS. Back then, in the olden days, you had to use the keyboard to do everything. That is a hard habit to break. As you have more than likely noticed the old menu bars are not there anymore. Before you break down and the tears start to roll, let's see what we can do.

Microsoft gave us some options for the keyboard user. Although the menus are not there, you can use the keyboard to access the different parts of the Ribbon. Not only can you access the Ribbon, but the old shortcuts (using the CTRL button) that you are use to using are still there and still working.

Below are tables showing some of the available keyboard shortcuts. A complete list of all of the shortcuts is provided for you on your computer in the Help section. To find the list click on the Help button (the small question mark in the upper right corner of the screen) and type keyboard shortcuts in the search window. Some of that list has reproduced for you in the tables below.

To do this	Press this
Delete one character to the left of the insertion point	BACKSPACE key
Delete one word to the left of the insertion point	CTRL+BACKSPACE (actually press Ctrl and then press Backspace and then release both keys)
Delete one character to the right of the insertion point	DELETE key
Delete one word to the right of the insertion point	CTRL+DELETE (actually press Ctrl and then press Delete and then release both keys)

Open the Office Clipboard	Press ALT+H to move to the **Home** tab, and then press F,O.
Copy selected text or graphics to the Office Clipboard	CTRL+C (actually press Ctrl and then press C and then release both keys)
Cut selected text or graphics to the Office Clipboard.	CTRL+X (actually press Ctrl and then press X and then release both keys)
Paste the most recent addition to the Office Clipboard.	CTRL+V(actually press Ctrl and then press V and then release both keys)
Move text or graphics once.	F2 (then move the cursor and press ENTER)
Copy text or graphics once.	SHIFT+F2 (then move the cursor and press ENTER)
Copy the header or footer used in the previous section of the document.	ALT+SHIFT+R (actually press the Alt key then press the Shift key and then press R then release all three keys)
Make letters bold.	CTRL+B (actually press Ctrl and then press C and then release both keys)
Make letters italic.	CTRL+I (actually press Ctrl and then press C and then release both keys)
Make letters underline.	CTRL+U (actually press Ctrl and then press C and then release both keys)

Decrease font size one value.	CTRL+SHIFT+< (actually press Ctrl then press Shift then press < and then release all three keys)
Increase font size one value.	CTRL+SHIFT+> (actually press Ctrl then press Shift then press > and then release all three keys)
Decrease font size 1 point.	CTRL+[(actually press Ctrl and then press [and then release both keys)
Increase font size 1 point.	CTRL+] (actually press Ctrl and then press] and then release both keys)
Remove paragraph or character formatting.	CTRL+SPACEBAR (actually press Ctrl and then press the Spacebar and then release both keys)
Paste special	CTRL+ALT+V (actually press Ctrl then press Alt then press V and then release all three keys)
Paste formatting only	CTRL+SHIFT+V (actually press Ctrl then press Shift then press V and then release all three keys)
Undo the last action.	CTRL+Z (actually press Ctrl and then press Z and then release both keys)

Redo the last action.	CTRL+Y (actually press Ctrl and then press Y and then release both keys)
Open the **Word Count** dialog box.	CTRL+SHIFT+G (actually press Ctrl then press Shift then press G and then release all three keys)

Table 1-1

To Insert this	**Press this**
A field	CTRL+F9 (actually press Ctrl and then press the F9 key and then release both keys)
A line break	SHIFT+ENTER (actually press Shift and then press Enter and then release both keys)
A page break	CTRL+ENTER (actually press Ctrl and then press Enter and then release both keys)
A column break	CTRL+SHIFT+ENTER (actually press Ctrl then press Shift then press Enter and then release all three keys)
The copyright symbol	ALT+CTRL+C (actually press Alt then press Ctrl then press C and then release all three keys)
The registered trademark symbol	ALT+CTRL+R (actually press Alt then press Ctrl then press R and then release all three keys)

The trademark symbol	ALT+CTRL+T (actually press Alt then press Ctrl then press T and then release all three keys)

Table 1-2

To Move	**Press This**
One character to the left	LEFT ARROW Key
One character to the right	RIGHT ARROW Key
One word to the left	CTRL+LEFT ARROW (actually press Ctrl and then press the Left Arrow key and then release both keys)
One word to the right	CTRL+RIGHT ARROW (actually press Ctrl and then press the Right Arrow key and then release both keys)
One paragraph up	CTRL+UP ARROW (actually press Ctrl and then press the Up Arrow key and then release both keys)
One paragraph down	CTRL+DOWN ARROW (actually press Ctrl and then press the Down Arrow key and then release both keys)
One cell to the left (in a table)	SHIFT+TAB (actually press Shift and then press the Tab key and then release both keys)
One cell to the right (in a table)	TAB Key

Up one line	UP ARROW Key
Down one line	DOWN ARROW Key
To the end of a line	END Key
To the beginning of a line	HOME Key
Up one screen (scrolling)	PAGE UP Key
Down one screen (scrolling)	PAGE DOWN Key
To the end of a document	CTRL+END (actually press Ctrl and then press the End key and then release both keys)
To the beginning of a document	CTRL+HOME (actually press Ctrl and then press the Home key and then release both keys)

Table 1-3

To Do This	Press This
Switch a paragraph between centered and left-aligned.	CTRL+E (actually press Ctrl and then press E and then release both keys)
Switch a paragraph between justified and left-aligned.	CTRL+J (actually press Ctrl and then press J and then release both keys)
Switch a paragraph between right-aligned and left-aligned.	CTRL+R (actually press Ctrl and then press R and then release both keys)
Left align a paragraph.	CTRL+L (actually press Ctrl and then press L and then release both keys)
Indent a paragraph from the left.	CTRL+M (actually press Ctrl and then press M and then release both keys)
Remove a paragraph indent from the left.	CTRL+SHIFT+M (actually press Ctrl and then press Shift and then press M and then release all three keys)
Remove paragraph formatting.	CTRL+Q (actually press Ctrl and then press Q and then release both keys)

Get Help or visit Microsoft Office Online.	F1 Key
Move text or graphics.	F2 Key
Repeat the last action.	F4 Key
Choose the **Go To** command (**Home** tab).	F5 Key
Go to the next pane or frame.	F6 Key
Choose the **Spelling** command (**Review** tab).	F7 Key
Extend a selection.	F8 Key
Update the selected fields.	F9 Key
Show KeyTips.	F10 Key
Go to the next field.	F11 Key
Choose the Save As command.	F12 Key
Start context-sensitive Help or reveal formatting.	SHIFT+F1 Key
Copy text.	SHIFT+F2 Key
Change the case of letters.	SHIFT+F3 Key
Repeat a **Find** or **Go To** action.	SHIFT+F4 Key

Move to the last change.	SHIFT+F5 Key
Go to the previous pane or frame (after pressing F6).	SHIFT+F6 Key
Choose the **Thesaurus** command (**Review** tab, **Proofing** group).	SHIFT+F7 Key
Shrink a selection.	SHIFT+F8 Key
Switch between a field code and its result.	SHIFT+F9 Key
Display a shortcut menu.	SHIFT+F10 Key
Go to the previous field.	SHIFT+F11 Key
Choose the Save command (Microsoft Office Button).	SHIFT+F12 Key
Choose the Print Preview command (Microsoft Office Button).	CTRL+F2 Key
Close the window.	CTRL+F4 Key
Go to the next window.	CTRL+F6 Key
Insert an empty field.	CTRL+F9 Key
Maximize the document window.	CTRL+F10 Key
Lock a field.	CTRL+F11 Key

Choose the Open command (Microsoft Office Button 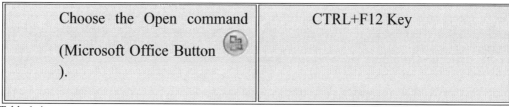).	CTRL+F12 Key

Table 1-4

As you can see there are several pages of shortcut keys available for you to use. This is not even a complete list of all of the shortcut keys that are available. If you still need more commands, click on the help button and search for Keyboard Shortcuts. The Help button is the small question box at the top right side of the screen.

Are you ready for even more good news? Microsoft has included new shortcuts with the Ribbon. Why you might ask. It is because this change brings two major advantages. First there are shortcuts for every Tab on the Ribbon and second because the many of the shortcuts require fewer keys.

The new shortcuts also have a new name: **Key Tips**

Using the keyboard press the Alt key

Pressing the Alt key will cause the **Key Tip Badges** to appear for all Ribbon tabs, the Quick Access Toolbar commands, and the Microsoft Office Button. After the Key Tip Badges appear, you can press the corresponding letter or number on the badge for the tab or the command you want to use. As an example, if you pressed Alt and then H you would bring the Home tab to the front. Figure 1-21 shows what the Ribbon looks like after pressing the Alt key.

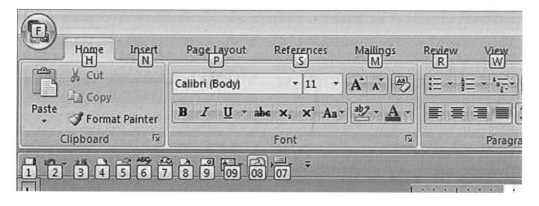

Figure 1-21

Note: You can still use the old Alt + shortcuts that accessed the menus and commands in the previous versions of Word, but because the old menus are not available, you will have no screen reminders of what letters to press, so you will need to know the full shortcut to be able to use them.

But that is not all of the shortcuts available. Microsoft has included shortcut menus. These are menus that you can access by right-clicking on an object or text. The shortcut menu is shown in Figure 1-22.

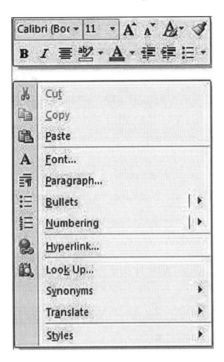

Figure 1-22

This shortcut menu shows all of the things that are available for you to do with the selected object. The bottom part looks a lot like the Edit from the previous version's menu bar. You will also notice the different formatting options that you can perform from the shortcut menu. See I told you that this version was cool!

Lesson 1 – 7   Creating a Document

When you first start Word 2007 you should have a blank screen like the one shown in Figure 1-23.

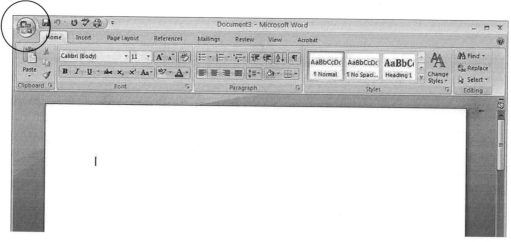

Figure 1-23

From here you can immediately start typing to create a new document. For this lesson we are going to start as if there was not a fresh clean slate to work with. Using the screen on your computer as the starting point, we will create another new blank screen. This is the technique that you will use if there is not a blank document area on your computer screen.

Click the Office Button

The Office Menu will appear and is partially shown below.

Figure 1-24

Click on New

A dialog box will appear on the screen with Blank Document highlighted.

Click on the Create Button

A new blank document will come onto the screen. The insertion point (The flashing vertical line) should be in the upper right corner of the main document area.

Note: if you have the new document icon in the Quick Access Toolbar and you click on it, you will get the same results as you did by performing the above steps.

Note: When you type the following memo, if there is a blank line separating the paragraphs, press the enter key one time for each blank line. Do not press the enter key to separate the lines inside of the paragraph. Word will automatically continue the text on the next line. This is called Word Wrap.

Type the following memo

To all committee members:  Press Enter twice here

The second Monday of the month is reserved for our regular committee meeting. The regular meeting time is 7:00 p.m. Please mark your planners and set your computer planner to reflect this.

Bill Mills – Committee Chair   Press Enter here

That is all there is to creating a new document. That was pretty easy, wasn't it?

This will bring us into our next lesson, how to save the document.

Lesson 1 – 8   Saving the Document

Now that you have gone through all of the trouble of entering the text, we need to save it so that nothing will happen to our memo. I don't know about you, but I have gone through situations like this only to have the power go off and everything is gone. Before that happens, lets learn how to save the Document.

Click the Office Button

When you click the Office button a drop down menu will appear as shown in Figure 1-25.

Figure 1-25

Note: The recent documents side of yours will not look like the one in figure 1-25; since these are my recent documents not yours.

You will notice that there are two save choices. Now would be a good time to explain the difference between the two of them. If this is the first time you have saved the current document, The Save choice will bring the Save As dialog box to the screen. If this is not the first time you have saved the document the choice will have different results.

Save: This will replace the existing document with the newer version that is displayed on your screen. All changes made will be saved and the original document, before you made any changes, will be gone.

Save As: This will allow you to save the currently displayed document with a different name. This will allow you to keep the original document just as it was before any changes were made to it and the new version will also be saved only under a different name.

Move the mouse to the Save As choice

Another menu will slide out to the side.

The first thing you have to do is make an important decision: what format will you use to save this document. Figure 1-26 shows the choices you have to choose from.

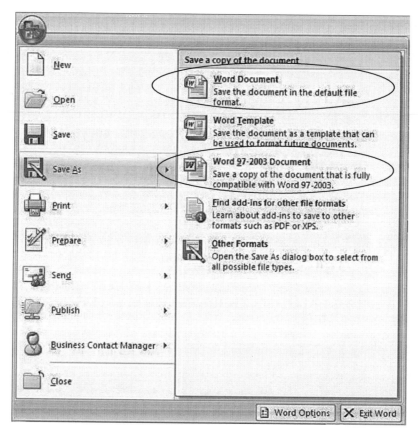

Figure 1-26

The two most obvious choices are: Save in the default format, which is Word 2007 and Save in the older format of Word 97 – 2003. If you were going to share this with someone who does not have Word 2007 on their computer but has an older version of Word, you would choose the Word 97-2003 document choice.

Choose the top choice Word document

This will bring the "Save As" dialog box to the screen. Again, if this is the first time you have saved this document the "Save As" dialog box will also come to the screen. The reason for this is because the first time you save a document you have to give it a name. If it already has a name and you select the Save choice it will save the new version. I had you choose the Save As choice just to make sure you knew about the different versions available. If you had simply clicked the save choice, Word would have saved the document in the default format which is Word 2007.

The "Save As" dialog box appears asking where to save the document and what name you want to give the document as shown in Figure 1-27. The default location is usually My Documents (unless you are already working in another folder), and the suggested name (usually the first line of the document) are already in place. If they are acceptable, simply click on "Save" to save the document. Your "Save As" screen may look different because these are my documents not yours.

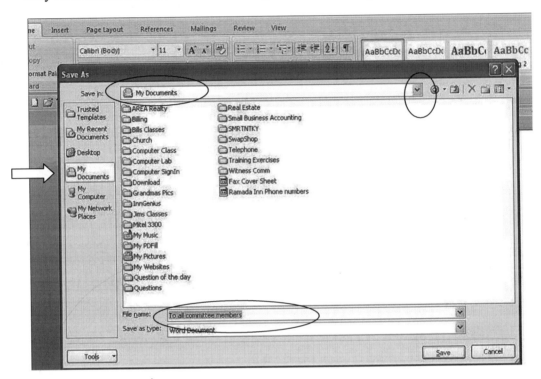

Figure 1-27

Make sure My Documents is selected on the right side (if it is not click on it with the mouse).

Now all we have to do is give the document a unique name. It is essential that every file has a unique name. This will allow us to keep our computer organized and also allow us to find the file when we need it. The name should reflect something about the document, such as why it was created. This document is being used for a memo to all of our committee members, so a name like "To All Committee Members" is more practical than a name like Document 17.

Click the Save button

...will disappear from the screen and the document will be saved as a Word 2007 file called To All Committee Members. You will be able to access this file any time you want by locating it under My Documents and double clicking it with the mouse. We will discuss opening an existing document in Lesson 1-10.

That is all there is to saving a file. Remember use Save As if you wish to keep the original document as it was before any changes were made and use Save if you wish to replace the original file with the revised version.

Note: From now on, when you are asked to save a document, make sure you save it in the MY Documents folder until you are told differently.

Lesson 1 – 9   Closing the Document

Now that we have saved our work we can safely close the file. This will be a very short lesson, as there is not very much to closing a document (file).

Click on the Office Button

The Office Button drop down menu is shown in Figure 1-28.

Figure 1-28

Click on the Close choice at the bottom

The document will close and you will end up with the main part of the screen being blank. In the next lesson we will see how to open the document.

Lesson 1 – 10   Opening a Document

If necessary open Word

Remember that Word is located in the Microsoft Office folder under all programs which is under the Start button.

There are three basic ways to open an existing Word Document. We will go through each method in this lesson. We will look at what is probably the most common way first.

Click the Office Button

The Office drop down menu is shown in Figure 1-29.

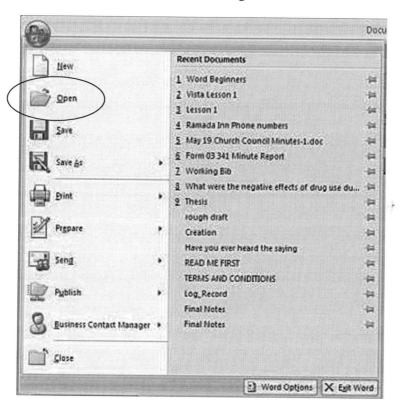

Figure 1-29

Click the Open Button

This will bring the Open Dialog Box to the screen as shown in Figure 1-30.

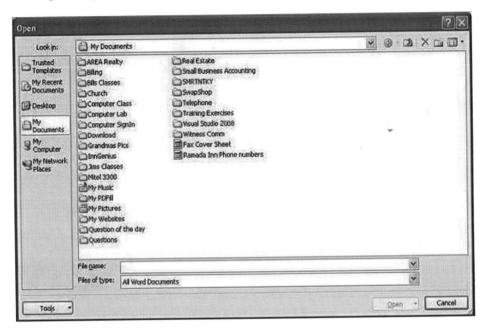

Figure 1-30

The normal starting place to look for a file is under My Documents, so that is the default place that should come up when you first see the Open dialog box. Since this is the place where we saved the file, you should see a file named To All Committee Members.docx. You will notice that the Open button is faded out and you cannot click on it at this point. Right now the button is disabled. As soon as you click on a file name the Open button will change and be enabled and be available to be clicked.

Click on To All Committee Members **and then click the Open button**

As soon as you click the Open button, Word will start opening the document. It may take a second or two depending on the speed of your computer for the screen to actually change. But in a few moments the document will whisk onto your computer screen and you will be able to work with it.

Close the document as we did in lesson 1-9

The second method for opening a document may be a little easier, but I am not sure that as many people use it as the first method we used.

Click on the Open button on the Quick Access Toolbar

The Quick Access Toolbar is show in Figure 1-31 and the open button is circled.

Figure 1-31

This will cause the Open Dialog box to immediately jump to the screen and you can continue just as we did on the previous page.

Click the Cancel button so nothing will open at this time

The third method for opening a file is about as easy as it gets if you have recently had the file open.

Click the Office Button to open the drop down menu as you did at the beginning of this lesson

This time all we need to do is click on the name of the file on the right that we want to open. See Figure 1-32 to help explain this.

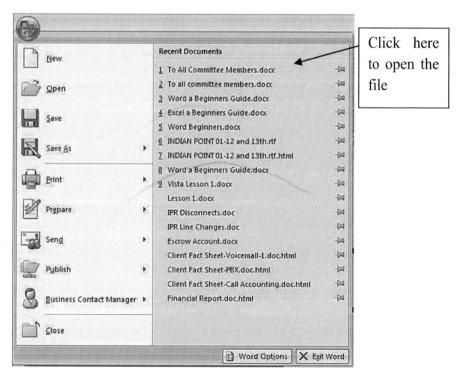

Figure 1-32

Lesson 1 – 11   Printing a Document

Let's recap: We can enter text, save the file, close the file, and open a file. Now let's see about printing a document. The document may be part of a report you are submitting or it may just be your personal letter. Either way you will probably want a printed copy. Printing the document is a simple process, provided that you have a printer set up on your computer.

If it is not open, open the document named To All Committee Members

The document should be displayed on the screen and look like it did when we last worked on it.

Click the Office Button and move the mouse to the Print choice, but do not click the mouse

There are three choices you can make from this menu: Print, Quick Print, and Print Preview. Before we actually print the document, it might be a good idea to see how it will look when we print it. We may find out that it will not all fit on one page and we may have to do some adjusting to our sheet. The available print choices can be seen In Figure 1-33.

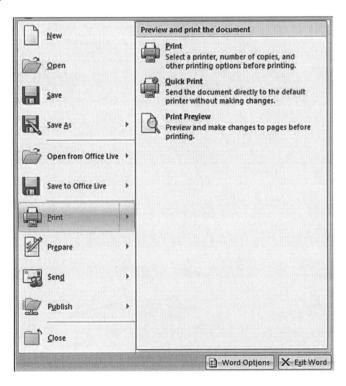

Figure 1-33

Click on the Print Preview choice

This will display how the sheet will look when it is printed. There are also a few other options available from this screen (see Figure 1-34).

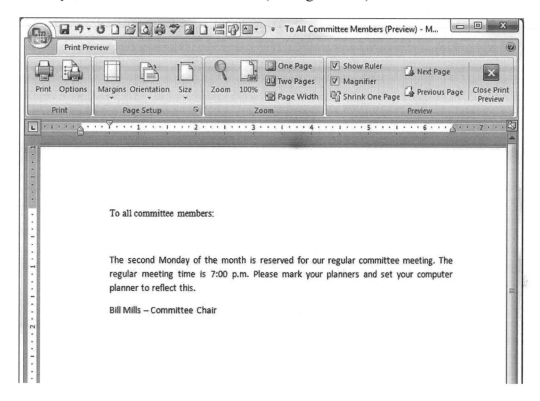

Figure 1-34

From this screen you can: Close the print preview, view the other pages (if there are any other pages), Zoom in or out on the document, make page settings, or print the document.

In the Page Setup group click the Margins command

This will show you the preset margins that are available to use. These are shown in Figure 1-35.

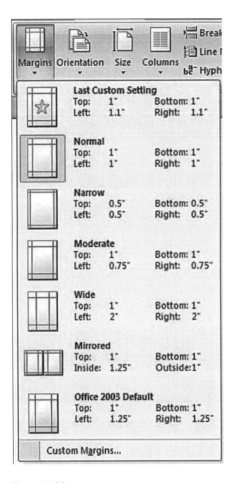

Figure 1-35

As you can see, this particular document is using the Normal margins, which is the default for all documents. If you want to change the margins away from the default, simply click on the choice and the dialog box will go away and the margins will be set to whatever you clicked.

If you do not see a margin that will work for this document and you would like to make a custom margin, click on the Custom Margins choice.

Click on Custom Margins

The Page Setup Dialog box will jump onto the screen and have the margins tab selected. This is shown in Figure 1-36.

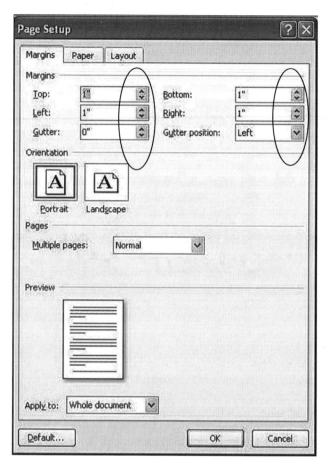

Figure 1-36

Using this dialog box you can manually adjust the margins to fit your needs. The top, bottom, right, and left margins are easy to understand. You may not, however, be familiar with the gutter. The gutter is an extra space on the side or top to allow for binding the document. This will prevent some of the text form being obscured when the document is bound.

Click the Cancel button

Now we will look at the Zoom feature. This feature will allow us increase or decrease the magnification on the Print Preview screen. It will not change the actual document.

Click the Zoom button

This will bring the Zoom Dialog box to the screen as shown in Figure 1-37.

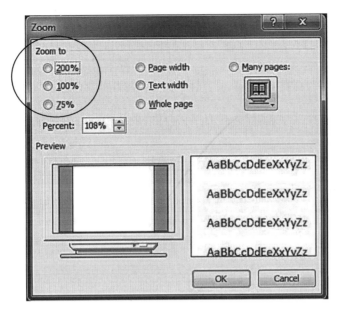

Figure 1-37

There are preset choices that you can zoom to. There is also a percentage box you can use to manually adjust the amount of zoom.

Click the radio button next to 200% and then click the OK button

This will make the print considerably larger in the Print Preview and a lot easier to read.

Click on the 100% button in the Zoom Group and see the difference it makes

Suppose that your document needs to be printed so that the long side is across the top instead of on the side? Changing the orientation will allow us to accomplish just that.

Click on the Orientation button

A short drop down menu will appear. This is shown in Figure 1-38

Figure 1-38

There are two choices in this menu, Portrait and Landscape. The Portrait choice is the normal way a paper is printed with the long side on the right and left sides. The Landscape choice prints with the long sides of the paper across the top and bottom.

Click on the Landscape choice

As you can see the print preview has changed to show you what the document will look like with this choice.

Change the orientation back to Portrait

The next problem you might run into is what if the document is just a little too long to fit on one page?

You may want to try adjusting the paper length. This can be done by changing from letter size paper to legal size paper. It should go without saying, but here it is anyway. To print on legal size paper you will have to change the settings and you will need to actually change the paper in the printer to legal size paper.

Click on the Size button

Another menu will drop down showing the different sizes of paper available. This is shown in Figure 1-39.

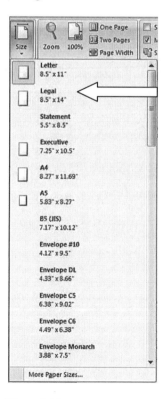

Figure 1-39

45

You will be able to see that letter is the default size of paper for the printer. You can change this by clicking on the desired paper size.

Click on Legal size

The size of the paper in the print preview area is now considerably longer. You may have to use the scroll bar on the right side of the screen to view all of the paper that is below the printed area.

Change the paper size back to Letter

The Preview group on the Print Preview screen has not been discussed yet. It is shown in Figure 1-40.

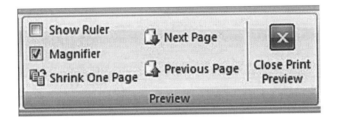

Figure 1-40

With this group you can view the next page, if there is a next page and the previous page if you are not on page 1. Clicking the checkbox next to Show Ruler will turn the ruler seen at the top and along the left side of the screen of and on.

If you click the Magnifier, the mouse pointer will turn into a small magnifying glass. Clicking the page with the mouse pointer will magnify that section of the page. Clicking the page a second time will turn the page back to the original size. The magnifier will remain on until you uncheck the checkbox to turn it off.

Clicking the Shrink One Page button will tell Word to try to shrink the document by one page by reducing the size and spacing of the text.

In the Print Group we will find an icon of a printer. Clicking this button will bring the Print Dialog box to the screen. This is shown in Figure 1-41.

Click the Print button

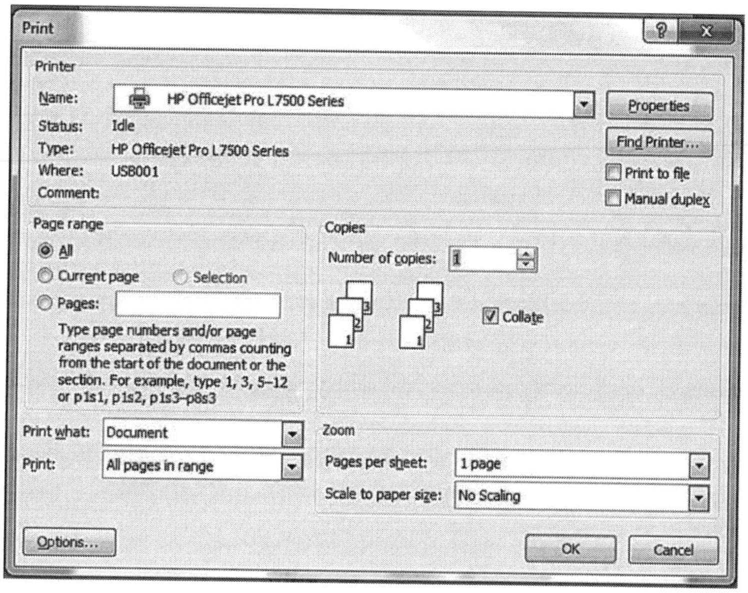

Figure 1-41

The Print Dialog box will allow us to choose the printer we wish to use by clicking the downward pointing arrow across from the word Name, assuming, of course, that there is more than one printer configured on your computer.

You can also choose to print all of the pages in the document, or the current page, or you can specify which pages to print. You can also chose to collate the pages (put them in the correct order as they print). If we click OK the document will print. If we click Cancel the Print Dialog box will go away and the document will not print.

Click the Cancel button

Click the Close Print Preview button

Now that we have seen what it is going to look like when we print it, let's see what the other two choices are when we moved the mouse over the print option.

Click on the Office button and move the mouse down to the word Print and then click on the top choice: Print

This will bring the same Print Dialog box to the screen as was shown in Figure 1-41.

In addition to the above mentioned choices we can also choose to print only the selection that we have highlighted. This choice will become available if we actually have some text highlighted. On the right side we can choose how many copies we want to print.

On the bottom left, we can choose to print all of the pages in the document or just the odd or even numbered pages. We can also choose to scale the document to fit a different size of paper.

Once we have everything marked correctly, all we have to do is click on the OK button and this will send the information to the printer.

Click Cancel so nothing will print

The last choice is the Quick Print choice and if you click on this choice the entire document will be sent to the default printer. There will be no dialog box for you to make choices. The document will just be sent to the printer.

Note: If you click on the print choice from the Office button, the Print Dialog box will come onto the screen. This is identical to the top choice of the three available choices we just discussed.

Chapter One   Review

The Word screen is divided into the following parts:

At the top: The Office Button, The Quick Access Toolbar, The Title Bar, and The Ribbon.

In the center: The Main Document Area, The Insertion Point, and the Scroll Bar.

At the bottom: The Status Bar and the Zoom slider.

Clicking the Office Button will bring a drop down menu to the screen where you can create a new document, open an existing document, save a document, print a document, prepare, send or publish a document. You can also link a document to the communications history of a business record. You can also close a document and exit Word from this menu. From this menu you can also access the options for Word.

The Ribbon consists of Tabs, Groups, and Commands.

Tabs:
>There are seven basic tabs across the top.
>The Home Tab contains the commands that you use most often.
>The Insert Tab contains all of the objects that can be inserted into a document.
>The Page Layout Tab contains the choices for how each page will look.
>The References Tab contains things such as Table of Contents and Footnotes.
>The Mailings Tab contains such things as Starting a Mail Merge and creating envelopes and labels.
>The Review Tab has things that are related to proofing, protecting, and comments.
>The View Tab allows you to change to the different views that are available.

Groups:
>Each Tab has several Groups that show related items together.
>Look at the Home Tab to see an example of the related Groups.
>The Home Tab has the following Groups: Clipboard, Font, Paragraph, Styles, and Editing.

Commands:
>A Command is a button, a box to enter information, or a menu.

Some groups have more commands than are visible. These commands are normally found in the associated Dialog Box. You open the Dialog Box by clicking the mouse on the Dialog Box Launcher located in the lower right corner of the group.

The Quick Access Toolbar is designed to give you easy access to the commands that you use over and over when working with Word. You can add commands to the Quick Access Toolbar by clicking the down arrow at the end of the toolbar and then clicking the desired command. If the desired command is not in the standard commands you can click on the More Commands choice and then add the command from the various choices.

Pressing the Alt key on the keyboard will bring the Key Tip Badges to the screen. These are shortcuts to access the Tabs of the Ribbon as well as the Office Button and the commands on the Quick Access Toolbar. The shortcuts using the Ctrl and Alt keys are still available, but you have to know the keystrokes to use them.

A new document can be created by clicking the Office Button and choosing "New" and then Blank Document and then click on the Create button. A new document can also be created by clicking the New Document command on the Quick Access Toolbar, if you added the command to the toolbar.

You can save a document by clicking the Save or Save as choice from the Office Menu, or by clicking the Save command on the Quick Access Toolbar, if you added the command to the toolbar.

Documents can be printed by choosing the Print option from the Office Menu or by clicking the Quick Print command on the Quick Access Toolbar, if you added the command to the toolbar.

Chapter One   Quiz

1) The Ribbon is divided into Task Oriented _____.

2) The Office Drop Down Menu is divided into two sections. The lesser used options are placed where?

3) Which section of the Word Options Dialog Box will allow you to add items to the Quick Access Toolbar?

4) On the Ribbon, a _____ is a button, a box to enter information, or a menu.

5) Print Preview is a command that can be added to the Quick Access Toolbar from the Standard Choices. **True or False**

6) What two keys on the keyboard, when pressed, will make the selected text italic?

7) What key, on the keyboard, do you press to cause the Key Tip Badges to appear?

8) If this is not the first time you have saved this document and you choose the Save choice from the Office Menu, will the Save as Dialog Box come to the screen. **Yes or No**

9) The Office Menu has a Print choice. If you move your mouse to the Print choice, what three commands are available?

10) On the Page Setup Dialog Box and on the Margins Tab, what is the Gutter?

Chapter Two   The Ribbon – A Closer Look

In chapter one, we had an overview of the Ribbon. In this chapter we will take a closer look at the different parts of the Ribbon.

In this chapter we will look at each standard tab of the Ribbon and each group that is on the Tab.

A word of concern: Depending on the size of your monitor, you may not see everything that is shown in the figures. Some of the groups may be condensed and may not show all of the available options at all times. The options are still available, but you may have to click one of the drop-down arrows to see them. I will show you some example of this as we continue.

This chapter will not have much user interface and consists of mostly reading (sorry about that), but it is necessary to have an understanding of the Ribbon.

Note: There are a few other tabs on the Ribbon that are not listed in the standard tabs. These only become available when they can be used. One example is if you insert a picture into your document the Picture Tools will appear with the format tab. These will be discussed when they are available.

Note: As we go through the rest of this book, the various groups and commands will be discussed as we use them.

Note: As we look at the different tabs in the book, click on each tab as we discuss it. This will allow you to see how it looks on your computer.

Lesson 2 – 1   The Home Tab

Open the document named To All Committee Members, **if it is not open**

The first thing I will show you is two different views of the same Home tab. One will show the view using a large monitor and the second is with a smaller monitor. You will be able to see the difference between the two and perhaps understand what I was saying on the previous page.

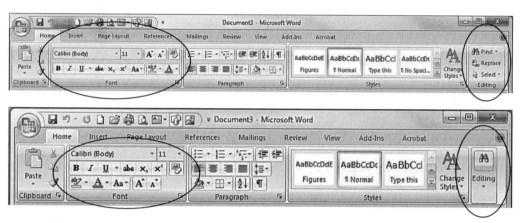

Figure 2-1

If the monitor is smaller the Ribbon will try to compensate by being taller when it is not as wide. Also some of the commands in the different groups may not be visible. Two examples of these are the Font group and the Editing group.

As I stated earlier the Home tab contains the commands that one normally uses most often when using Word 2007.

Let's take a closer look at the different groups on the Home tab.

The first group (on the far left side) is the Clipboard Group and is shown in Figure 2-2.

Figure 2-2

This group deals with the different things you can do with selected text, and objects.

The next Group is the Font Group and is shown in Figure 2-3.

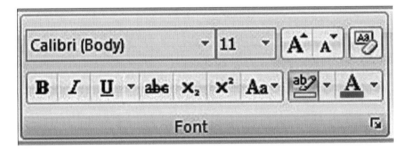

Figure 2-3

As you would expect, this is where you would perform all of the formatting for the text.

The next group is the Paragraph Group and is shown in Figure 2-4.

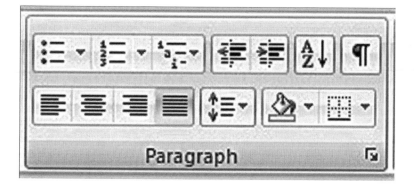

Figure 2-4

This group contains the formatting for the paragraphs.

The next group is the Styles Group and is shown in Figure 2-5.

Figure 2-5

This group contains the options for applying and setting styles to the document.

The last Group on the Home Tab is the Editing Group and is shown in Figure 2-6.

Figure 2-6

This group deals with searching for and replacing text.

Lesson 2 – 2   The Insert Tab

The Insert Tab, as you might expect deals with the various things you can insert into your document.

The first group on the Insert Tab is the Pages Group and is shown in Figure 2-7.

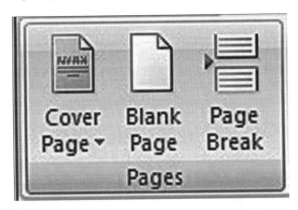

Figure 2-7

This group deals with inserting a Cover Page, a blank page, and a Page Break.

The second group is the Tables Group and is shown in Figure 2-8.

Figure 2-8

This deals with inserting tables into your document.

The next group is the Illustrations Group and is shown in Figure 2-9.

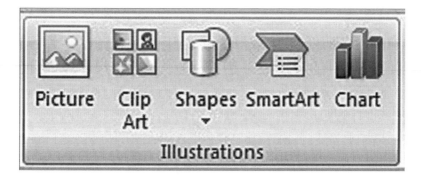

Figure 2-9

This deals with inserting pictures and charts into your document.

The next group is the Links Group and is shown in Figure 2-10.

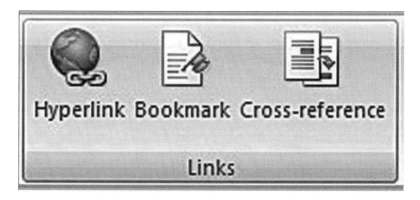

Figure 2-10

Using this group you can put in a link to another document or to a place inside the current document.

The next group is the Header & Footer Group and is shown in Figure 2-11.

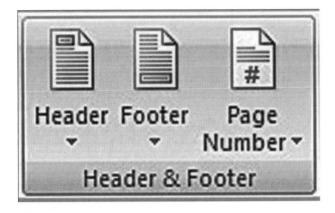

Figure 2-11

This group will allow you to insert page numbers as well as insert and edit headers and footers in your document.

The next group is the Text Group. This group is shown in Figure 2-12.

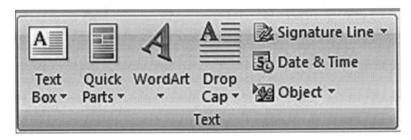

Figure 2-12

This group deals with the various text objects that you can insert into your document.

The last group on the Insert Tab is the Symbols Group and is shown in Figure 2-13.

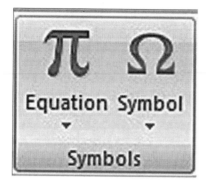

Figure 2-13

This deals with inserting mathematical calculations and special symbols into your document.

Lesson 2 – 3   The Page Layout Tab

This tab deals with the way your document is laid out on the screen and on paper when it is printed.

The first group is the Themes Group and is shown in Figure 2-14.

Figure 2-14

This group deals with changing the overall design and layout of the document.

The second group is the Page Setup Group and is shown in Figure 2-15.

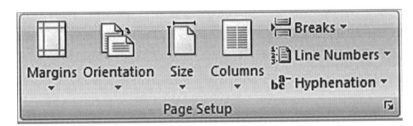

Figure 2-15

This group deals with such things as margins and paper size.

The next group is the Page Background Group and is shown in Figure 2-16.

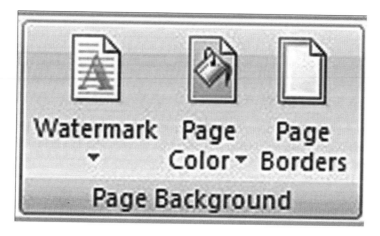

Figure 2-16

This will allow you to make changes to the background of the page in your document.

The next group is the Paragraph Group and is shown in Figure 2-17.

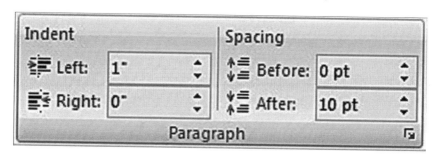

Figure 2-17

This group will allow you to work with indentations and Spacing.

The Arrange Group is the last group on the Page Layout Tab and is shown in Figure 2-18.

Figure 2-18

This group works with the relationship between text and objects, such as pictures.

Lesson 2 – 4   The References Tab

The first group is the Table of Contents Group and is shown in Figure 2-19.

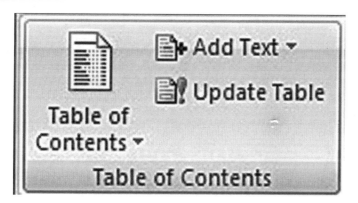

Figure 2-19

The group allows you to add and/or edit the Table of Contents in your document.

The second group is the Footnotes Group and is shown in Figure 2-20.

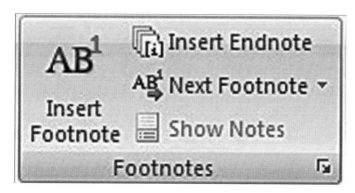

Figure 2-20

This group works with adding footnotes to your documents.

The next group is the Citations and Bibliography Group and is shown in Figure 2-21.

Figure 2-21

This allows you to cite another book or article as a source of information for something in your document.

The next group is the Captions Group and is shown in Figure 2-22.

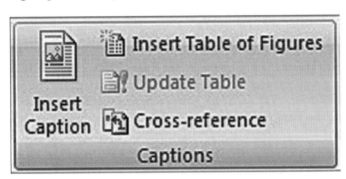

Figure 2-22

This group deals with inserting captions under pictures and inserting a table of all of the Figures in the document.

The next group is the Index Group and is shown in Figure 2-23.

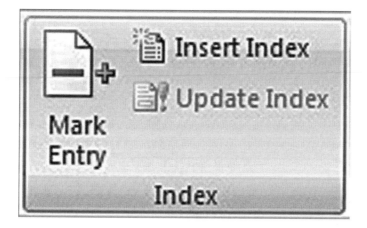

Figure 2-23

This group would be used if you were inserting an index into your document.

The last group on the References Tab is the Table of Authorities Group and is shown in Figure 2-24.

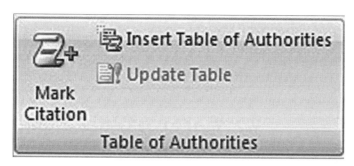

Figure 2-24

The Table of Authorities is a list of cases, statutes, and other authorities used in your document.

Lesson 2 – 5   The Mailings Tab

The Mailing Tab deals mainly with using the Mail Merge Feature in Word. Chapter Ten is devoted to the Mail Merge feature.

The first group on this tab is the Create Group and is shown in Figure 2-25.

Figure 2-25

This group deals with creating envelopes and labels.

The second group is the Start Mail Merge Group and is shown in Figure 2-26.

Figure 2-26

This group is where you will start the Mail Merge process.

The next group is the Write and Insert Fields Group and is shown in Figure 2-27.

Figure 2-27

This group deals with inserting fields into the Mail Merge.

The next group is the Preview Results Group and is shown in Figure 2-28

Figure 2-28

This will allow you to see how the documents will look before you print them.

The next group is the Finish Group and the last group is the Marketing Group. They are shown in Figure 2-29.

Figure 2-29

This is where you would finish the Mail Merge and set up a new marketing campaign.

Lesson 2 – 6   The Review Tab

This tab deals with such things as proofing your document, adding comments, and tracking changes to your document.

The first group is the Proofing Group and is shown in Figure 2-30.

Figure 2-30

This group is used mainly when you are editing your document or adding text to the document.

The second group is the Comments Group and is shown in Figure 2-31.

Figure 2-31

This group is where you add comments to your document.

The next group is the Tracking Group and is shown in Figure 2-32.

Figure 2-32

This group is for keeping track of changes to your document.

The next group is the Changes Group and is shown in Figure 2-33.

Figure 2-33

This group also deals with changes that were made to your document. From here you can decide if you want to accept the changes or reject them.

The last two groups are the Compare and Protect Groups and they are shown in Figure 2-34.

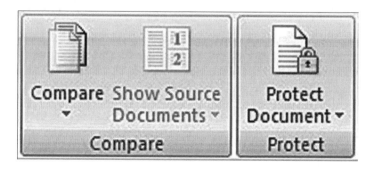

Figure 2-34

The Compare Group deals with comparing multiple versions of a document and the Protect Group deals with protecting your document from someone accessing it and making changes to it.

Lesson 2 – 7   The View Tab

The View Tab mainly deals with the different ways that you can view your document.

The first group on the View Tab is the Document Views Group and is shown in Figure 2-35.

Figure 2-35

This group allows you to view your document in several different layouts.

The next group is the Show/Hide Group and is shown in Figure 2-36.

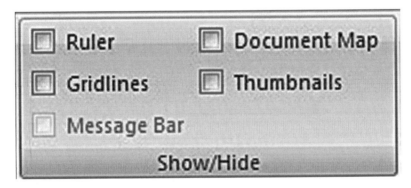

Figure 2-36

This group allows you to either show or hide various things from view.

The next group is the Zoom Group and is shown in Figure 2-37.

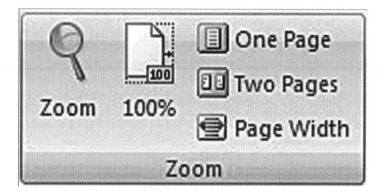

Figure 2-37

The Zoom Group will allow you to magnify what you are looking at on the screen.

The next group is the Window Group and is shown in Figure 2-38.

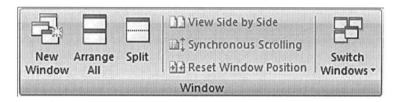

Figure 2-38

This deals with how you can view windows if you have more than one window open.

The last group is the Macros Group and is shown in Figure 2-39.

Figure 2-39

Macros come into play if you have to type the same text over and over again. This will allow you to type it once and place it where ever you need it.

Chapter Two   Review

The Home Tab contains the following Groups:

    Clipboard

    Font

    Paragraph

    Styles

    Editing

The Insert Tab contains the following Groups:

    Pages

    Tables

    Illustrations

    Links

    Header and Footer

    Text

    Symbols

The Page Layout Tab contains the following Groups:

    Themes

    Page Setup

    Page Background

    Paragraph

    Arrange

The References Tab contains the following Groups:

    Table of Contents

    Footnotes

    Citations & Bibliography

    Captions

    Index

    Table of Authorities

The Mailings Tab contains the following Groups:

    Create

    Start Mail Merge

    Write & Insert Fields

    Preview Results

    Finish

The Review Tab contains the following Groups

      Proofing

      Comments

      Tracking

      Changes

      Compare

      Protect

The View Tab contains the following Groups:

      Document Views

      Show / Hide

      Zoom

      Window

      Macros

Chapter Two   Quiz

1) What tab would you use to add a hyperlink to your document?
2) What tab would you use to add a macro to your document?
3) What tab would you use to add a chart to your document?
4) What tab would you use to change the font formatting?
5) What tab has the copy and paste command on it?
6) What tab would you use to add comments to your document?
7) What tab will hide the Ruler at the top?
8) What tab will allow you to insert an index into your document?
9) What tab will allow you to work with the Page Background?
10) What tab will allow you to add a cover page to your document?

Chapter Three   Editing Text

Almost without exception, you will write a letter, memo, or document of some kind and find a better way of communicating your thoughts. It may turn out that you need to move text to another place in the document or you may need to add or delete certain text. This chapter will deal with the various things available when you are editing your documents.

Lesson 3 – 1   Selecting Text

When it comes to editing text in a document, the first thing you have to understand is how to select text. Before you can do anything with text you must first select it. This lesson will show you how to select text.

Open the document titled To All Committee Members **if it is not open**

We will use the short memo that we created earlier for this lesson.

There are several ways to select text, so we will start with the most common.

The most common method of selecting text is by moving the mouse pointer to the text, clicking and holding down the left mouse button and dragging the mouse over the text, and then releasing the mouse button. Let's give it a try.

Select the text "To all committee members:"

To select the text, move the mouse pointer until it is just before the T in "To all committee members". Click and hold the left mouse button down and drag the mouse to the right until the rest of the line is highlighted. Now release the left mouse button.

If you did this correctly the document should look like Figure 3-1.

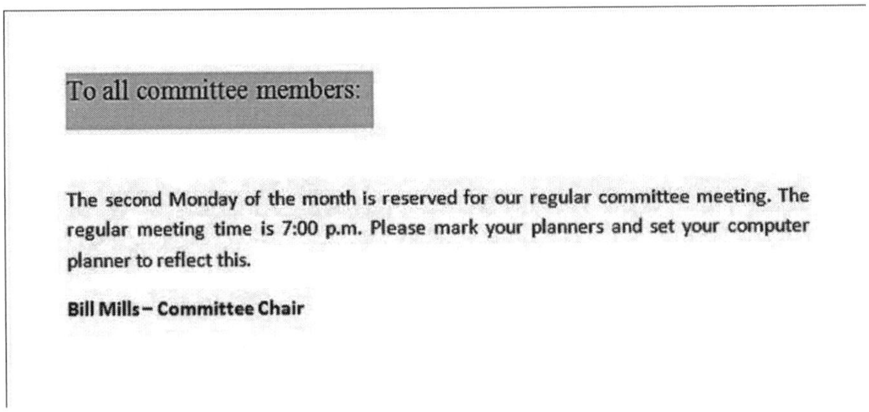

To all committee members:

The second Monday of the month is reserved for our regular committee meeting. The regular meeting time is 7:00 p.m. Please mark your planners and set your computer planner to reflect this.

Bill Mills – Committee Chair

Figure 3-1

As you can see, when the text is selected it shows up as being highlighted.

This can be a little tricky, so you may have to try it a few times before it works.

Text can be de-selected by moving the mouse anywhere and then clicking and releasing the left mouse button.

This is not the only way text can be selected. Text can also be selected by clicking on it.

Click the mouse on the word Monday **in the first paragraph**

The insertion point will just sit there flashing inside the word. This is just what it is suppose to do when you click the mouse once.

With the mouse pointer inside the word, double-click the mouse

Monday should now be highlighted.

Click anywhere outside of the highlighted word to unselect it

Now you know two ways to select text, but these are not the only way you can select text.

Move the mouse to the left margin of the document, across from "To All Committee Members", until it changes into a small white arrow

Click the mouse once

The entire line is highlighted. This can be handy if you need to make changes to the line, such as deleting the entire line.

Click somewhere off the line to unselect it

Move the mouse back to the left border, across from the main body of the memo, and double-click the mouse

This time the entire paragraph is highlighted.

Unselect the paragraph and then move the mouse back to the left border and triple-click the mouse

This time the entire document is selected. For you keyboard users, this is the same as holding down the Ctrl key and pressing the A key.

Selecting text was easy and you will use this constantly when you are editing a document.

Lesson 3 – 2   Using Undo and Redo

Before we continue on in our discussion of Editing, there is one other thing that will come into play: Using the Undo and Redo feature.

Being human, sometimes we make mistakes. If I had not been forced to use this feature over and over again during the writing of this book, it probably would not seem as important as it actually is. Microsoft, in their infinite wisdom, looked into the future and knew that I was going to use their product and added this feature probably just for me. I will explain it to you just in the off chance you may need it.

Guess what? Undo is not on the Ribbon. Just when you thought the Ribbon held everything you would ever need, it doesn't have the Undo button.

The Quick Access Toolbar is shown in Figure 3-2, and is the home of the Undo and Redo buttons.

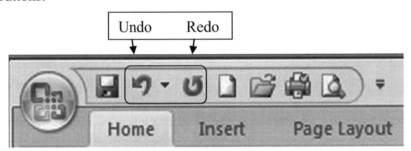

Figure 3-2

Any time you copy, type, cut, paste, or do almost anything the Undo button becomes available. If you make a mistake, you can click the Undo button and everything will be as it was before you made the mistake. Now let's be realistic here, Word will not know that you didn't really want to do the silly thing that you just did. That means that you can't make the mistake today and tomorrow when you realize that you made the mistake, expect Word to undo it. If you click the Undo button, Word will undo the last thing that you did, not the mistake you made five minutes ago. Obviously this may not be magic, but it is close.

Open the document To All Committee Members

Select the line To: All Committee Members **and press the Delete key**

This line is now removed from the document, it disappeared. Oh No, you didn't really want to delete that line! We can have Undo correct the mistake.

Click the Undo button on the Quick Access Toolbar

The line "To: All Committee Members" is now back into the document.

Suppose that after careful consideration you really did want to delete the line. Do we have to re-select it and delete it again? No, Microsoft wouldn't do that to us. There is another button that will let us redo the delete. This button is right next to the Undo button and is called the Redo button.

Click the Redo button

The line is now gone again. As I said it may not be magic, but it is close.

Lesson 3 – 3   Insert and Delete Text

Inserting and deleting text is something that you will do almost every time you use Word, so we might as well practice it. This lesson will show you how to do just that.

Open the To All Committee Members **document if necessary**

Click your mouse immediately after the period at the end of the first paragraph, the main body of the memo, and press Enter.

This will start a new paragraph.

Type the following and press Enter when you are finished.

> The July meeting will not be on this date, as it is my birthday. The meeting for July will be held on the third Monday.

That is all there is to inserting text. It really is that easy. Let's try it again.

Click your mouse at the end of the line Bill Mills – Committee Chair **and press Enter.**

This will start another new paragraph.

Type the following and press Enter when you are finished

> P.S. Presents for the chair on his birthday will not be considered as sucking-up, just common courtesy.

See it really is that easy.

There will be times when you decide that something you put into a document should not be there and you will need to delete it. Word has made this so easy you will not lose any sleep over how to get this done.

Select the last paragraph of the memo (Let's try a new way to select the paragraph).

Click the mouse on the word birthday in the paragraph.

This will move the insertion point inside the paragraph. Now comes the fun part. If you double click the mouse, you will select the word birthday, but if you triple click the mouse, you will select the entire paragraph (you can also select the entire paragraph by double-clicking in the left margin of the paragraph).

Triple click the mouse

This will select the entire paragraph.

On the keyboard, press the Delete key

The entire paragraph is now gone. Deleting and inserting text are some of the easiest things you will do in Word.

Bring the paragraph back by using the Undo button

Save the changes and close the document

Lesson 3 – 4   Using Cut and Paste

This lesson and the next lesson are dedicated to moving text and objects around in your Document. We will be using Cut, Copy, and Paste to accomplish this. First, let me give you a brief description of these commands.

Cut: This will <u>remove</u> any data that is highlighted (selected) and move it to the clipboard. The clipboard is a temporary storage place and will temporarily hold the data until you can put it someplace else in the document or even into a completely different document or program.

Copy: The copy command will <u>make a copy</u> of the selected text (or object) and place it in the clipboard for you to use later.

Paste: The paste command will <u>copy from the clipboard</u> and put the information into the document.

Once you have selected the text, you can move it to another place in the document by cutting and then pasting it elsewhere. Cutting and pasting text is one of the most common things you will do in Word. When you cut text, it is removed from its original location and put it in a temporary storage area called the Clipboard. You can then move the insertion point to a new location and paste the text from the Clipboard. The Clipboard is available from any program in Windows, so you can cut text from one program and paste it into another program.

Toward the beginning of this book there is a page that has Important Notice across the top. This page tells you how to download the files referenced in this book. If you have not downloaded these files, do so now.

Open the document titled BranTel Info.docx

This document is located in the files that you downloaded from the website, which you probably put in My Documents.

On the second page of the document we want to change the order of two of the paragraphs. We can use cut and paste for this.

Select the second paragraph on the second page, the one that deals with certification, and the blank line below the paragraph

When this is selected it will be highlighted in blue (See Figure 3-3). After you have selected the text it can be cut from the document.

Support:
>Support will be provided to the dealer 24 hours a day seven days a week. If everyone is busy calls will go to voicemail and the technician will return calls as soon as they are free. A toll free number will be provided to the dealers for tech support. Certification will be required to talk to tech support. At present there are no plans to charge for technical support.

Certification:
>Training and certification will be provided, and required, at our facilities in Branson. This will be provided at no charge to the dealer (travel, lodging and meals not included). Classes will cover installation, changing or adding boards, features, general use of the system, programming and troubleshooting.

Payment:
>The first shipment will be sent COD certified check or company check. Master Card or Visa will be accepted over the phone. After a credit application has been returned and approved all accounts are net 20 days.

Figure 3-3

Click on the Cut command

As you remember, the cut command is on the Home Tab of the Ribbon and in the Clipboard Group. If your monitor is not large enough to show the word cut, the command is identified by a pair of scissors. As soon as you click the cut button, the text disappears from the document. The text is not really gone; it has been placed on the clipboard and is now ready for us to paste it to a new location.

Click the mouse at the beginning of the word Support

Click the Paste button

The text will magically jump back into the document where the insertion was located.

Note: You can also use the right-click shortcut to cut and paste the text. If the text is highlighted and you right-click on it, the following shortcut menu will appear on the screen.

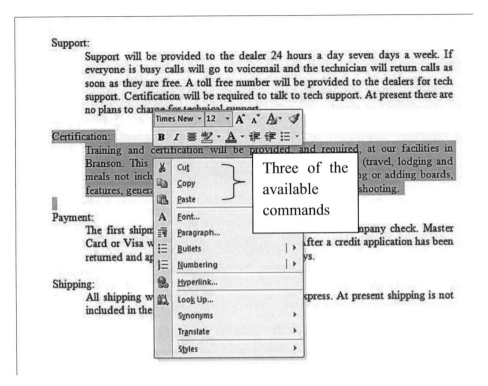

Figure 3-4

As you can see, along with the formatting commands are the cut, copy, and paste commands ready for you to use.

Save your changes but do not close the file

Remember to use the Save As choice and save the document in the My Documents folder. You can keep the same name, but make sure it is the My Documents folder. This will leave the original file unchanged.

Lesson 3 – 5   Viewing Documents Side-By-Side

and

Using Copy and Paste

There may be a time when you need to compare two documents and check for differences, or similarities. You could go back and forth between the documents and check all of the text, or you could view both documents at the same time. In this lesson we will compare two documents on one screen and copy text between them.

Open the BranTel Info2 **document**

This document is also with the downloaded files. The task bar at the bottom of the screen should look like Figure 3-5.

Figure 3-5

You could switch between the documents by clicking on the left box and then on the right box, or you could have both documents on the screen at the same time. This way you could compare the documents side-by-side.

Click on the BranTel Info **document icon in the Task Bar**

This will make sure this document is on the left side of the screen and your screen will look like the figures.

Click on the View Tab of the Ribbon and select View Side by Side in the Window Group

This will allow both documents to be on the screen at the same time, sharing the screen equally. The result of clicking this button is shown in Figure 3-6.

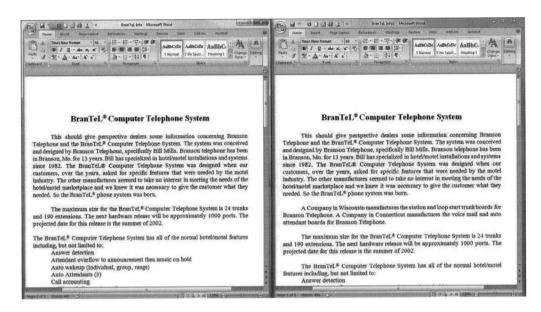

Figure 3-6

As you compare the document, you will notice that the one on the right has a paragraph that is not in the document on the left. We will now copy the paragraph from the document on the right and paste it in the document on the left.

Click inside the document on the right and then select the second paragraph (see Figure 3-7).

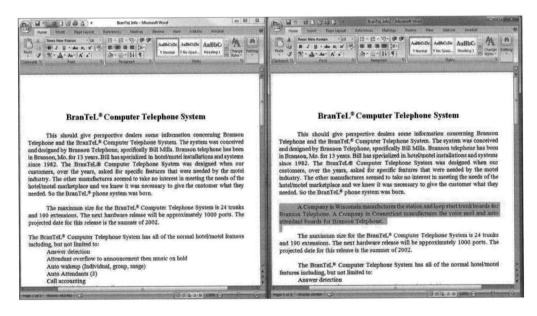

Figure 3-7

Click on the Home Tab of the document on the right and select Copy

The copy command is just below the cut command and is represented by the icon that looks like two pieces of paper.

Click on the document on the left and then click just in front of the first word at the beginning of the second paragraph

Click on the Home tab of the document on the left and choose the Paste command

Since we used the Copy command instead of the cut command, the paragraph remained in the document on the right. The paragraph was pasted into the document on the left and is now in both documents.

The documents should look like Figure 3-8.

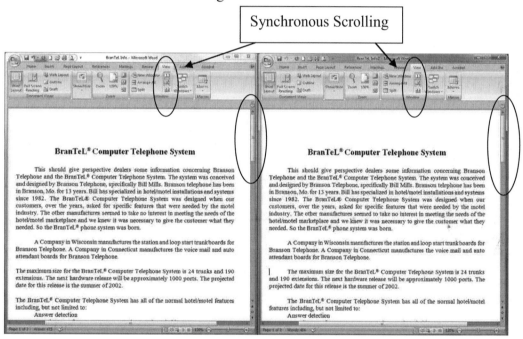

Figure 3-8

The View Side by Side button has changed its appearance and is now in a different color, as shown in Figure 3-8. If you want to go back to the original views, click the Side by Side View button.

Just below this button is another button that is also a different color. This is the Synchronous Scrolling button. This ensures that if you scroll down in one document the other document scrolls down at the same time and at the same pace. This can be turned off and on by clicking this button.

Click the Synchronous Scrolling button if it is not an amber color

Using one of the scrolling bars slowly move down the document to the second page

Both documents should move together as you scroll down the screen. Using this you can compare the documents quite easily.

Remove the Side by Side view by clicking the View Side by Side button

Close both documents and save any changes made

Remember to use the Save As choice and save the document in the My Documents folder. You can keep the same name, but make sure it is the My Documents folder. This will leave the original file unchanged.

Lesson 3 – 6   Navigating through a Document

As a document gets longer, it gets harder and harder to navigate through it. Suppose you were typing a 200 page novel, how would you get to the very end of the document or simply to page 73? This lesson will show you how to move around inside a document.

Open the document named Testimonial **from the downloaded files**

One way to move around in a document is by using the Windows Scroll bars. The Vertical scroll bar is located on the right side of the window and is used to move up and down in a document. If there is a horizontal Scroll Bar, it is located on the bottom of the window and is used to move from left to right when a document is too wide to fit on one screen.

Click the down arrow on the Vertical scroll bar a few times (the arrow is located at the bottom of the scroll bar).

Every time you click the down arrow, the screen scrolls down one line.

Click and hold the down arrow on the Vertical Scroll Bar

This causes the screen to move downward more rapidly.

Click on and drag the vertical scroll box to the top of the vertical scroll bar

This will take you back to the top of the document.

Using the Keyboard, press the End key

This will move the insertion point to the end of the current line.

Press the Home key on the keyboard

This will move the insertion point to the beginning of the current line.

Press the Ctrl key and hold it down. Now press the End key and then release both keys

This will move the insertion point to the very end of the document. You might also notice that the vertical scroll box has moved down near the end of the scroll bar, indicating your position in the document. Look at the Status bar at the bottom of the screen.

The Status bar indicates what page of the document you are currently on (See Figure 3-9).

Figure 3-9

Press the Page Up button on the keyboard

This will move up one screen, not necessarily one page but one screen.

Press the Page Down button on the keyboard

This will move down one screen, not necessarily one page but one screen.

Press Ctrl and Home

This will whisk you back to the beginning of the document.

The table below lists the most common Keyboard shortcuts for moving around in a document

Home	Moves the insertion point to the start of the line
End	Move the insertion point to the end of the line
Page Up	Moves Up one screen
Page Down	Moves down one screen
Ctrl + Home	Moves the insertion point to the beginning of the document
Ctrl + End	Moves the insertion point to the end of the document

Table 3-1

If you would like to move to a specific page in the document you can.

The Editing Group is on the right end of the Home Tab. This group is shown in Figure 3-10.

Figure 3-10

On the right side of the word Find is a downward pointing arrow.

Click the arrow next to Find and then click on Go To

This will bring the Find and Replace dialog box to the screen. As shown in Figure 3-11.

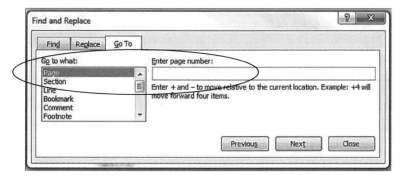

Figure 3-11

The Go To tab is in the front of the other tabs. In the Go To what section, Page is already highlighted and the insertion point is in the Enter Page Number textbox. All you have to do is type the page number that you want to bring to the screen and it will happen.

Using the keyboard type in the number 2 and press Enter

Immediately you will be taken to the beginning of page 2. You can use this to move to any page in the document.

Save and close this lesson

Remember to use the Save As choice and save the document in the My Documents folder. You can keep the same name, but make sure it is the My Documents folder. This will leave the original file unchanged.

Lesson 3 – 7   Using Drag and Drop

Back in Lesson 3–4 we learned how to cut text and paste it someplace else in our document. In this lesson we will learn an easier way to perform the same task. The method we are going to learn is called Drag and Drop. It is called this because we are going to select a section of text drag it from its current location and drop it someplace else in the document.

Open the To All Committee Members **document**

We will use this document to practice dragging and dropping text. I think you will enjoy moving text using this method. One reason is because it is so easy and the other reason is we all like having the power to command things to move.

Select the last paragraph

To drag the selected text we need to "grab" the text with the mouse.

If you move the mouse inside the highlighted area and then click and hold the left mouse button down, you will "grab" the selected text so that it can be moved. To move the text, we need to simply move the mouse and the selected text will follow.

There is something you need to watch for as you move the mouse. A small vertical dotted line will move along with the mouse. This will show you where the text will be inserted when you release the left mouse button. In Figure 3-12 you can see that the text will be inserted just before the word Bill if I release the left mouse button.

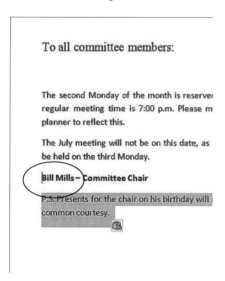

Figure 3-12

93

Using the mouse grab the selected text and drag it up and then drop it just before the word Bill

This may require a little practice but once you master this you can save a lot of time. If you did this correctly, your document should look like Figure 3-13.

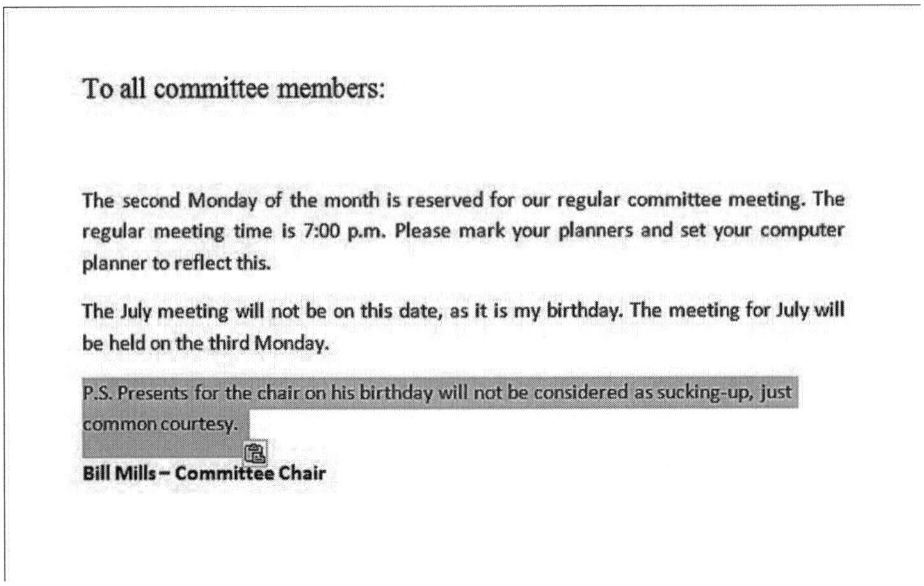

To all committee members:

The second Monday of the month is reserved for our regular committee meeting. The regular meeting time is 7:00 p.m. Please mark your planners and set your computer planner to reflect this.

The July meeting will not be on this date, as it is my birthday. The meeting for July will be held on the third Monday.

P.S. Presents for the chair on his birthday will not be considered as sucking-up, just common courtesy.

Bill Mills – Committee Chair

Figure 3-13

Undo the Drag and Drop you just made

Close the document

Lesson 3 – 8   Finding and Replacing Text

Searching for text in your document can be time consuming and frustrating. You know that somewhere in your document you referenced a certain word, but now you are trying to find it. Oh Boy! How much fun is this?

You could search through the entire document, word after word and page after page, or you could let Word do the search for you.

Open the Testimonial **document**

Click the Find Button

The Find button is on the right end of the Home Tab, in the Editing Group.

The Find and Replace Dialog box will come back onto the screen with the Find tab in the front. The Insertion point will already be positioned in the "Find what" text box ready for you to type in a word you want to find.

Using the keyboard type the word completely **and press the Enter key**

Word will search the document and stop on the word completely. The testimonial would sound better if it said extremely pleased instead of completely pleased. By the way, the dialog box will stay on the screen until you close it just in case you want to search for another word.

Click the Cancel button in the dialog box

Manually replace the word completely with the word extremely

You can also have Word automatically replace words for you.

Click on the Replace command

This command is just below the Find command in the Editing Group.

The same dialog box comes onto the screen with the Replace tab at the front (See Figure 3-14).

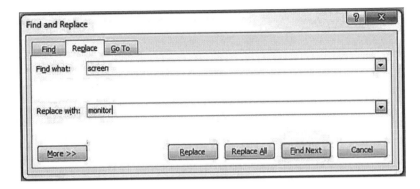

Figure 3-14

In this example we want to replace the word screen with the word monitor.

In the Find what textbox type the word screen

In the Replace with textbox type the word monitor

Now Word is ready to start searching your document to see if it can find the requested word.

Click the Find Next button

Starting from the insertion point Word will start searching for the word screen. If Word finds the word screen it will stop searching and stop on the word. At this point you can decide if this is the instance of the word screen that you want to replace. If it is, all you have to do is click on the Replace button.

Click the Replace button

The word screen has now been replaced with the word monitor. Now you can choose to find the next occurrence of the word screen or you can jump ahead and choose to replace all occurrences of the word.

A word of caution: If you click on the "Replace All" button, Word will replace every occurrence of the word screen with the word monitor; you will not get to decide which instance gets replaced. Word will automatically replace <u>all</u> occurrences of the word. Be very careful this may not really be what you wanted to accomplish.

Close the document without saving the changes

Lesson 3 – 9   Using Spell Check & Word Count

There is probably nothing more embarrassing than to have someone come up to you and correct your spelling and/or grammar in a document. Microsoft thought that this would be embarrassing for you as well and provided you with a spell check feature. Before you let anyone see your work, it is probably a good idea to run the spell checker.

Did you ever notice that when you are typing a word and press the space bar, a red line appears under the word? This is because Word automatically checks the spelling of words as you type them. Many people think this is the greatest feature of Word: the Spell Checker. Word identifies spelling errors, grammar errors, and repeated words as well. Word checks for these errors while you type, highlighting spelling errors with a red underline, and grammar errors with a green underline.

Word can also correct common mistakes for you as you type. If you typed hte Word could automatically change it to the.

For this lesson we will use a fictitious document titled Drugs and War that a young student might turn in as an assignment.

Open the document Drugs and War **that is in the downloaded files**

There are several typing errors in this document. These errors include both spelling and grammar errors.

There is more than one way to perform the Spelling/Grammar check. First we will use the most common method.

On the Quick Access Toolbar there is a button you can click to activate the spelling and grammar check feature. The icon for this is the check mark with the ABC above it (See Figure 3-15).

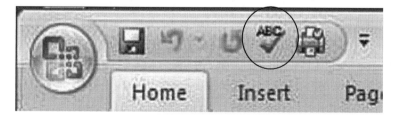

Figure 3-15

Make sure the insertion point is at the beginning of the document

Clicking the mouse at the beginning of the title will move the insertion point if it not already there.

Click the Spell Check icon on the Quick Access Toolbar

The Spelling and Grammar Dialog box will jump onto the screen as shown in Figure 3-16.

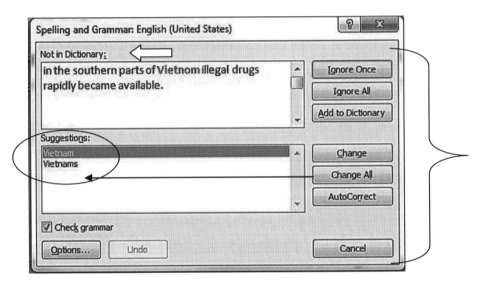

Figure 3-16

We are now at the first spelling error. There are several things you should notice about the dialog box. First: the word highlighted in Red is not in the dictionary. Second: Word is giving you suggestions as to what word might replace the highlighted word. Third: there are several choices on the right as to what you can do next.

You can choose to ignore this word once, or ignore all occurrences of this word. The obvious question is why would you want to ignore a misspelled word? One possibility is that some companies intentionally misspell a word as part of their logo or product name. As examples; some trucking companies have "Xtra" and "Xpress" on the side of their trucks.

If the highlighted word is correctly spelled, and you are sure that it is correct, you can add the word to the dictionary. I have added my e-mail user name to the dictionary so that it will not show up as a misspelled word.

If the correct spelling is shown under the suggestions, you can highlight the correct spelling and then click on change to replace your misspelled word with the correctly spelled word. If you wish you can replace all occurrences of this word (the highlighted word) with the correctly spelled word.

You can also choose the auto correct option. This will allow Word to decide what to replace the misspelled word with. Word will replace the misspelled word with what it thinks is the best possible choice. Be careful with this option, it may not be your best choice.

If you want, you can also cancel the spell check by choosing the Cancel choice.

In our scenario the word is misspelled and needs to be replaced by the correctly spelled word Vietnam.

Click the Change button

Now that the spelling is correct for the sentence, Word will check the grammar. Word noticed that the first word in the sentence is not capitalized. Instead of being highlighted in red, this error is highlighted in green. The green indicates that this is a grammar error instead of a spelling error. You will notice that in the suggestions the first letter of the word "in" is capitalized. This is what is needed in the sentence so we should click the Change button.

Click the Change button

The change will happen and the spell check feature will move to the next error.

This is the most common way to use the spell check feature, but not the only way. Let's try another way.

Click the Cancel button to close the spell check dialog box

We can still see the spelling and grammar errors in the document. They are still underlined in green or red. We can use this indication to help us make the corrections needed.

Right-click on the word selers

The shortcut menu will jump onto the screen with the options that are available. This menu is shown in Figure 3-17.

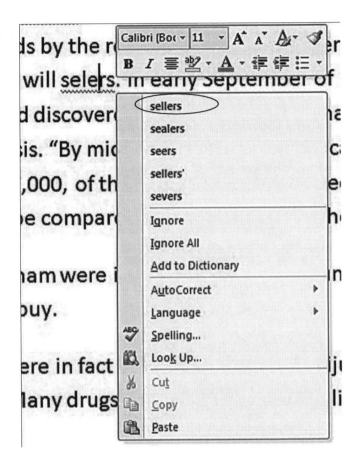

Figure 3-17

You can see that the same choices that were in the spell check dialog box are available from the shortcut menu.

Click on the correctly spelled word "sellers"

The misspelled word has been replaced and you are ready to move on to the next error.

Correct the remaining errors that Word found using one of the two mentioned methods

Please note the wording of the last thing that you were told to do. You were told to correct the remaining errors that Word found in the document. This would imply that there are errors that Word did not find. Reread the second sentence of the first paragraph. The way it was written doesn't make a lot of sense. All of the words are spelled correctly and Word did not pick up on this error.

The moral of this story is that the Spell Check is a great tool but it will not replace you actually proof reading the document yourself.

Replace the word will in the second sentence with the correct word with

Now that the Spell Check is finished, let's take a quick look at the Word Count feature. This was mentioned earlier in lesson 3-6 when we talked about the Status Bar. Now we will look at the Word Count command on the Ribbon. It is found on the Review Tab in the Proofing Group (See Figure 3-18).

Figure 3-18

Click on the Word Count command

The following information box will come onto the screen.

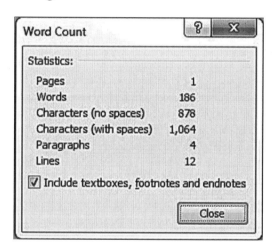

Figure 3-19

As you can see there is a lot of information in this little box. These include the number of pages, words, characters (with and without spaces), paragraphs, and lines. You can see that this gives you a lot more details than the Status bar did in the earlier lesson.

If you want information on just a section of the document instead of the entire document, you can highlight a section and then click the Word Count command.

Click the Close button

Highlight the first paragraph and then click the Word Count Button

The information you receive has changed to reflect only the highlighted selection not the entire document (See Figure 3-20).

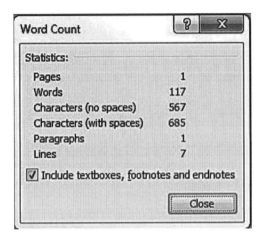

Figure 3-20

The information on the Status Bar has also changed to reflect the highlighted selection (See Figure 3-21).

Figure 3-21

The Status Bar indicates that the selected text has 117 words out of 186 total words.

Close the Word Count box and the Document (save the changes made)

Remember to use the Save As choice and save the document in the My Documents folder. You can keep the same name, but make sure it is the My Documents folder. This will leave the original file unchanged.

Lesson 3 – 10   Using the Thesaurus

Now that you have edited your text and used the spell check, you probably think that you are done. Guess what? This chapter is only a little over half finished. There are still plenty of things that fall under editing. This lesson will deal with using the Thesaurus. Most of you have probably never used the Thesaurus and you may not even know what it is let alone why you will want to use it. The Thesaurus is used to find other words with a similar meaning to the word you have selected. In this lesson we will learn how to find these similar words and how to put them into our document.

Open the Testimonial **document**

In this letter we will find a few words that could easily be exchanged for perhaps better words.

Select the word wonderful **in the second line of the second paragraph**

You can double-click the word to select it or click and drag the mouse over the word.

Click the Thesaurus icon in the Proofing Group on the Review Tab

The icon is shown on Figure 3-22.

Figure 3-22

Clicking the Thesaurus icon will bring the Research pane to the screen. As you remember the task pane is on the right side of the screen. Figure 3-23 shows the task pane.

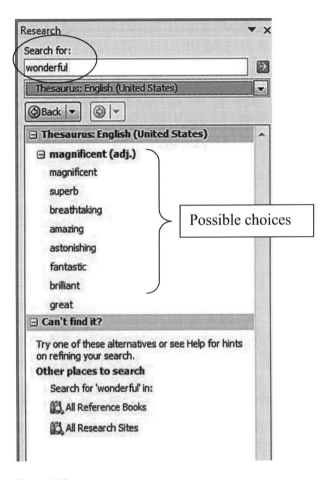

Figure 3-23

The task pane shows that we searched for the word wonderful and we searched the U.S. English Thesaurus. Below this are the words that we can substitute in place of wonderful. You will also notice that wonderful is being used as an adjective.

If we want to use one of these words, all we have to do is move the mouse over to the word and click on the down arrow that will appear and then choose insert from the menu. Figure 3-24 shows this.

Figure 3-24

Move the mouse over the word magnificent and click on the insert choice from the drop down menu

Magnificent will be inserted into your document and the word wonderful is now gone.

Save your changes and close the document

Lesson 3 – 11   Tracking Changes

At first I wasn't sure that I should include this lesson in the book. After careful consideration I decided that it was something to good to just pass over it. This feature is normally used if someone else is checking over your document and making suggestions for improvements or changes. Since it is so cool, I thought you should see it.

Open the Testimonial2 **document from the downloaded files**

Here are the ground rules for this lesson:

No one else in our company has the authority to make changes to the final document except you.

Before we send this document to someone to proof read it for us we need to keep tract of any changes that are made. To keep tract of the changes we need to turn the Track Changes Option on.

Click the Track Changes icon in the Tracking Group of the Review Tab (See Figure 3-25)

Figure 3-25

The Track Changes icon will turn to an amber color to show that it is active. When we send this document as an e-mail or if someone on the network works on it all changes will be tracked.

In this part of the lesson we will play the part of the person who is proof reading this document for someone else.

Double-click on the word completely toward the end of the first paragraph and then type the word extremely

You will notice that my change shows up in red and is also underlined in red. You will also notice that the word completely has a line through it.

Note: If we sent this to someone else and they opened the document on their computer their comments and changes would show up in a different color.

We also want to change the formatting of the second paragraph.

Select the second paragraph and click the Justify Alignment button on the Home tab

The Paragraph Group containing the alignment buttons are shown in Figure 3-26.

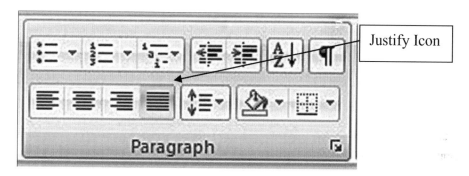

Figure 3-26

The screen will open up to show the Markup Area. This is an area that is not printed, but it is there for things like this. Figure 3-27 shows the Markup Area.

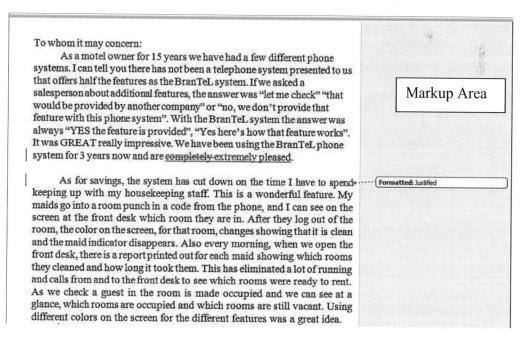

Figure 3-27

Let's make one more change. In the middle of the second paragraph there is a sentence that starts "Also every morning".

Select the word every **and change it to** in the

Close the document and save the changes when prompted

Reopen the document

Now we are going to switch roles. Now we are back to original writer of the document and we are ready to see the changes that were made to our letter.

The first thing that we will want to do is turn the tracking off. We do this because we don't want anything we do to show up as a change. When we are finished it will be the final document.

Click the Track Changes button on the Review Tab

The button will change back to its original color to show that it is turned off. Also any changes we make to the document will not be tracked. The document should look like Figure 3-28.

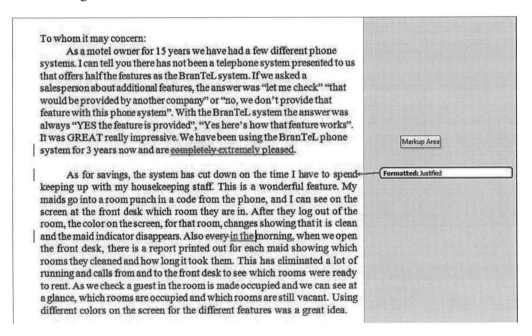

Figure 3-28

The vertical lines, on the left, show you the lines in the document that were changed. If text was added, it will be in a different color than the original text. Since there was only one person making changes, all of our changes are showing in red. First we have a suggestion to change one of the words.

Using the mouse, click inside the first change, the one where the word completely is changed to extremely

Now we have to decide if we want to accept this change or reject it. Ultimately if we accept it, the word with the strike through will be replaced with the word that was added by the person making the change. This is a two part process. First we have to decide if we want to accept deleting the word that has the line through it and second do we want to add the new word. Figure 3-29 shows the Accept and Reject buttons, which are located in the Changes Group just to the right of the Tracking Group.

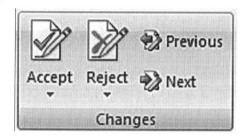

Figure 3-29

On the Review Tab click on the Accept button

The word completely is gone and the word extremely is highlighted waiting for us to make the decision about inserting this word into the document. If we click the accept button the word extremely will be inserted into the document. If we click reject the word extremely will be deleted.

Click the Accept button

The second paragraph will be highlighted and we can see that the formatting was changed to justify. If we like this all we have to do is click the Accept button.

Click the Accept button

The third change should automatically be highlighted and ready for us to either accept or reject the change. After some serious thought, we decide that we like the original wording and do not want to make this change.

Click the Reject button

The word "every" is no longer highlighted and will stay in the document. Now we have to decide if we still want to add the words "in the" to the letter. That would look really silly if we left it in the letter, so we need to reject this change also.

Click the Reject button

There are no more changes to the document and you will be informed of this (See Figure 3-30).

Figure 3-30

Click the OK button

I probably don't need to say this, but that has not stopped me yet. Before you send out the final version of your document, make sure you have addressed all of the changes and nothing looks out of place in your document. It would be embarrassing to send out a document that still has some underlined text and balloons on the side.

Close the document and save the changes

Remember to use the Save As choice and save the document in the My Documents folder. You can keep the same name, but make sure it is the My Documents folder. This will leave the original file unchanged.

Lesson 3 – 12   Inserting Symbols & Special Characters

There will be a time when you need to have a character in your document that you just can't seem to find a key on the keyboard that will let you put the character into your document. This lesson will show you how to find and insert these special characters and symbols.

Open the Testimonial **document**

In this document there is a reference to the BranTeL phone system. This is a registered trademark and needs to be identified as such.

Put the insertion point just after the "L" in BranTeL in the third line

The Symbols command is on the Insert tab of the Ribbon.

Click on the word Symbol

A menu will drop down showing some of the most recently used symbols, but this is not all of the symbols you can choose from.

Click the "More Symbols" choice at the bottom

Now you can really choose the symbols you need. Figure 3-31 shows the Symbol Dialog box.

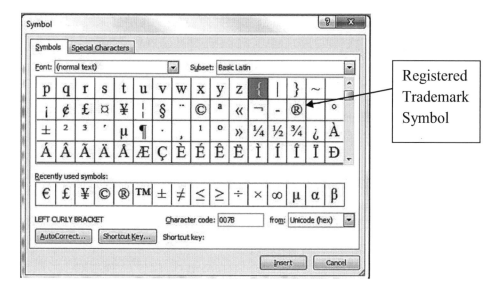

Figure 3-31

Navigate to and find the "Registered" symbol

Click on the symbol and then click the Insert button

Now the Registered Trademark symbol is right after the word BranTeL® just as it should be.

Note: If the symbol shows up as a regular size letter instead of the smaller size superscript letter, you will need to tell Word that this should be smaller and located above the line. Figure 3-32 will show you the difference between the two.

Figure 3-32

If this is the case, select the symbol and then click on the superscript icon in the Font group of the Home tab (See Figure 3-33).

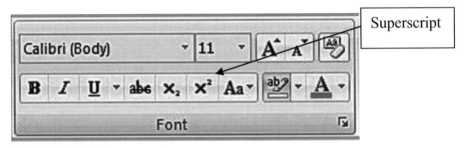

Figure 3-33

Finish adding the "Registered symbol" to the rest of the BranTeL occurrences

Obviously this is not the only symbol that you can add. You might decide to add a smiley face to your document. The process for adding any of the symbols is the same as adding this one.

Save your changes and close the document

Lesson 3 – 13   Adding Comments and Footnotes

Adding a comment to a document is quite easy and can be done in only a few moments. I guess the big question is why would you add a comment? I suppose that you could use the comment to remind yourself of something, perhaps something that you need to say for a presentation. First we will look at adding a comment to your document, and later we will look at adding footnotes.

Open the Teaching Lesson **document; also from the downloaded files**

In this document we outlined the steps needed to teach a new class. Now we will add comments that we will refer to as we go through the class.

It might surprise you to know that when you insert comments you do not use the Insert Tab. You add comments from the Review Tab. The Comments Group of the Review Tab is shown in Figure 3-34.

Figure 3-34

Where the comment is positioned in your document depends on where the insertion point is located at the time the comment is added.

Move the insertion point to the end of the first section immediately after the word lessons

Click the New Comment button

The result is shown in Figure 3-35.

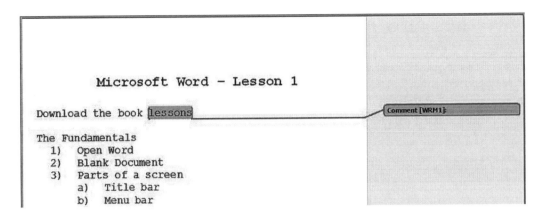

Figure 3-35

The word "lessons" is highlighted in red. This is because the insertion point was just after the word. This is the word that our comment is tied to. To the right is the actual comment area. First we have the word Comment followed by my initials and the number one. My initials are there because I am the author and the number one because it is the first comment. The insertion point is already in the comments box and is ready for you to start typing.

Type the following in the comment section

Make sure to save user name and password

Click anywhere outside the comments box

This will move you back to the document and you can finish working.

At the end of point three click the mouse after the word "Options"

Add another comments that says Plan on one hour for this section

Add another comment at the end of point eight that says Use handout here

Add another comment after point ten that says Page 28 in the book

Add another comment at the end of point eleven that says End on page 31

Your document should look like Figure 3-36.

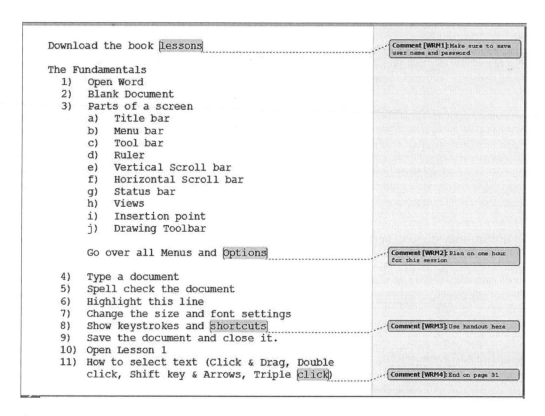

Figure 3-36

If you need to delete a comment, you can click on the comment and then click the Delete button that is next to the new comment button. The next and previous buttons allow you to move between the comments.

A footnote is slightly different than a comment. Footnotes are used in printed documents to explain, comment on, or provide references for text in a document. You might want to use footnotes for detailed comments. Footnotes are found at the bottom of the page and are numbered sequentially. You have probably seen these in most documents and manuscripts.

Let's add a quick footnote to the Teaching Lesson document.

The Footnotes Group is on the Reference Tab of the Ribbon and is shown in Figure 3-37.

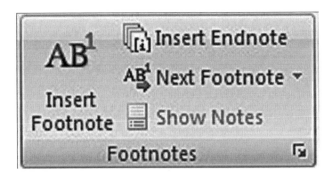

Figure 3-37

Click the mouse right after the word Fundamentals

Click the Insert Footnote command

This will immediately take you to the bottom of the page, where there is a small number 1. The insertion point is already there waiting for you.

Type the following:

These steps are taken from the book "Word a Beginners Guide".

When you are finished click anywhere inside the document

Go back to the top and look at the line The Fundamentals

You will notice that there is a small number 1 at the end of the line. This number represents the number of the footnote. If this was the second footnote, there would be a number 2 after the word Fundamentals.

As you can see both comments and footnotes are easily added to your documents. Not only are they easy, but your documents will look very professional.

If you have to cite something as a source of information, you could insert a citation instead of a footnote. A citation stays where you insert it and does not go down to the bottom as a footnote does.

Close the document and save your changes (Don't forget to use Save As)

Chapter Three   Review

The most common way of selecting text is to click and hold the left mouse button down just before the text that you want to select and drag the mouse to the end of the text that you want to select. Next you release the left mouse button. A single word can be selected by double-clicking the mouse while it is inside of the word. A complete line can be selected by clicking the mouse in the left margin directly across from the line you want to select. If you double-click the mouse while it is in the left margin, the entire paragraph will be selected. If you triple-click the mouse while it is in the left margin, the entire document will be selected.

The Undo and Redo buttons are located on the Quick Access Toolbar. Any time you copy, type, cut, paste, or do almost anything the Undo button becomes available. If you make a mistake, you can click the Undo button and the last thing you did will be undone.

To insert text all you have to do is click the mouse where you want the text and start typing. You can select text and press the Delete key on the keyboard to delete the text. One character directly to the right of the insertion point will be deleted if you press the Delete key and no text is selected. The Backspace key will delete one character to the left of the insertion point.

You can use cut, copy, and paste to move text around inside of the document.
Cut: This will remove any data that is highlighted (selected) and move it to the clipboard. The clipboard is a temporary storage place and will temporarily hold the data until you can put it someplace else in the document or even into a completely different document or program.
Copy: The copy command will make a copy of the selected text (or object) and place it in the clipboard for you to use later.
Paste: The paste command will copy from the clipboard and put the information into the document.

You can use the Side-by-Side view to view two documents at the same time on one screen. This is normally used when comparing two documents for similarities or differences. Synchronous Scrolling allows both documents to scroll down together if you click one of the scrolling bars.

Certain key(s) on the keyboard will help you move around inside of a document.

Home	Moves the insertion point to the start of the line
End	Move the insertion point to the end of the line
Page Up	Moves Up one screen
Page Down	Moves down one screen
Ctrl + Home	Moves the insertion point to the beginning of the document
Ctrl + End	Moves the insertion point to the end of the document

Table 3-2

The "Find" command will let you search for specific words in the document. The "Replace" command will let you search for a word and replace it with another word. The "Go To" command will let you move to a specific place in the document.

Using Drag and Drop you can move selected text to a different location inside the document.

Don't forget to use Spell Check on all of your documents.

Use the Thesaurus to find a word with a similar meaning,

If someone else is going to proof read your documents and then suggest changes, use the Track Changes command. Then you can decide to either accept the changes or reject them.

You can insert symbols and special characters into your documents.

You can also add comments and footnotes to your documents.

Chapter Three   Quiz

1) Double-clicking the mouse inside a paragraph will select the entire paragraph. **True or False**

2) Triple-clicking the mouse in the left margin of a page is the same as holding the Ctrl key down and pressing A key. **True or False**

3) Clicking the Undo command on the Quick Access Toolbar will cause Word to start searching for mistakes and then undo any mistakes it finds. **True or False**

4) The Copy command will make a copy of, and then delete the selected text. **True or False**

5) What command do you click to cause two documents, being viewed side-by-side, to scroll down at the same time?

6) What key(s), when pressed, will cause the Insertion Point to move to the beginning of the document?

7) Using Drag and Drop will let you move selected text to a new location inside the document with the mouse. **True or False**

8) If you use the Replace command and click the Replace All button, a dialog box will come to the screen for you to confirm the replacement at every occurrence of the word in the "Find What" section. **True or False**

9) The Word Count Information Box can tell you how many characters there are with and without spaces. **True or False**

10) The Thesaurus is used to find words with a similar meaning to the selected word. **True or False**

Chapter Four   Managing Files

This chapter is not really a part of Microsoft Office 2007. It is included in this book because it is so important. After you make your documents you will need to save them in a place that will enable you to find them easily when you need them.

If you are one of those people who have your entire monitor screen filled with icons and your documents folder is so full you have trouble finding the correct file, this chapter may change your computer life.

In this chapter we will learn how to make folders and how to move your documents around so that each document gets in the correct folder. You will learn how to organize your computer.

Just so that you remember, these documents we have been making are actually files that you are storing on your hard drive.

Lesson 4 – 1   Making New Folders

If you were at home storing papers you probably wouldn't put them in a big pile in the center of your table or desk. You would more than likely put them in a filing cabinet. In your filing cabinet you would not just randomly push papers into any drawer. You would have several folders to keep similar documents together, and have the papers all neatly stacked inside each folder. This is the same line of thought you need to have with your computer.

Before you can organize your files, you need a place to keep them. Our first lesson is dedicated to teaching you how to make folders.

Click on the Start button and select My Documents on the right side

The Windows Explorer program will jump onto the screen. On your screen you will have one of two possible views. The left side of the screen may have "My Documents" at the top and it will be highlighted (see Figure 4-1), or the left side may not list any of the folders at all (See Figure 4-2). Your screen may look different if you are using Vista or Windows 7.

Figure 4-1   Figure 4-2

The right side of the screen will show all of the files and folders that are inside the folder. The My Documents folder is an ideal place to save your files. However putting all of your files in the My Documents folder can also make this folder cluttered and make it hard to find things.

If you had other folders inside of the My Documents folder you could start to organize your computer.

Let's make another folder inside of the My Documents folder. If you have the view that is shown in figure 4-2 making a new folder is quite easy, you click on Make a new folder. If you have the view that is shown in figure 4-1 it is a little harder. Just to have everyone on the same page, if you have the view that is shown in figure 4-2 click on the folders icon that is shown in Figure 4-3.

Figure 4-3

We all have the same view and can work together.

Click on File in the menu bar at the top

Move the mouse to New and then click on Folder

Figure 4-4 shows this process.

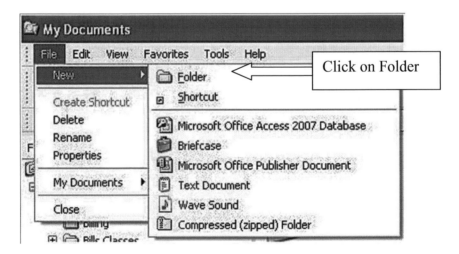

Figure 4-4

If you check your computer screen you will find a new folder underneath the My Documents folder. It is named New Folder (See Figure 4-5).

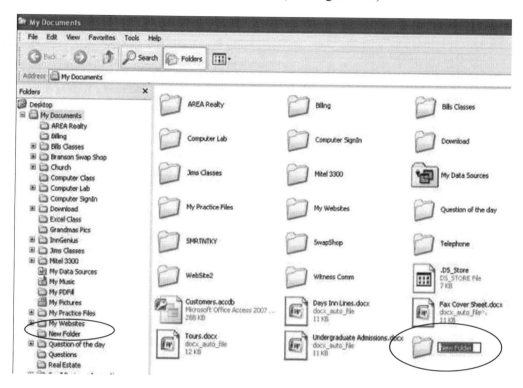

Figure 4-5

The actual folder on the right has the name highlighted and you are expected to give this folder a different name. If you are going to use this folder to store files it should have a name that reflects the files that are stored in it. Let's use this folder to store the files that we will make as we travel through this book.

Using the keyboard type Word 2007 **and then press the Enter key**

You do not have to click the mouse or anything, just start typing. You will notice that as you type the "New Folder" name is replaced with the name Word 2007.

Using this same technique you can create folders to keep all of your like files together. You may want one folder for family letters or that famous Christmas letter that you send out every year.

In the next lesson you will learn how to move files into your new folder.

Lesson 4 – 2   Moving Files

Let's put the documents we have been using inside the Word 2007 folder. In this lesson I will show you how to move a file (document) from one folder to another folder.

I asked you way back in lesson 1-8 to make sure you saved all of your documents in the My Documents folder. Now we are going to move them to the Word 2007 folder. If you didn't save them in the My Documents folder, I don't know where they will be, you are going to have to look for them. To find them you may have to click on Start and then search and then all files and folders, then you can enter the name of the document (such as To All Committee Members) and then click search. In the results section, you will be able to see where the file is located.

The rest of this lesson is going to assume that your documents are located in the My Documents folder.

Click on the Start button and then click on My Documents.

You should see, somewhere in there, a set of files that look something like the ones shown in Figure 4-6.

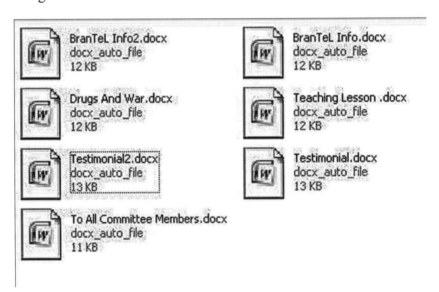

Figure 4-6

Click on the file named To All Committee Members and drag it over to the folder named Word 2007 and then release the mouse button

As you pass over the folders, each one will highlight to let you know that if you release the mouse button this is the folder your file will be moved to. When the folder named Word 2007 is highlighted, you can release the left mouse button. The file named To All Committee Members should now be gone from the My Documents folder. Let's see if it in the Word 2007 Folder.

Double-click on the Word 2007 folder to open it

The results should look like Figure 4-7.

Figure 4-7

Using this same procedure move the remaining documents to the Word 2007 folder

The documents are named BranTeL Info, BranTeL Info2, Testimonial, Testimonial2, Drugs and War, and Teaching Lesson.

That is all there is to moving files from one folder to another.

Armed with this knowledge, you can start making folders to keep similar documents together. Go ahead, start making folders and move your documents. Now you can start organizing your computer.

Lesson 4 – 3   Copying Files

Just for the sake of having something to do, we want to keep our documents in the Word 2007 folder, but we also want the "To All Committee Members" file to be in the My Documents folder. To accomplish this we will need to copy the file from one folder to another, not move it.

Double-click on the Word 2007 folder to open it if it is not already open

Your screen should look similar to the one shown in Figure 4-8.

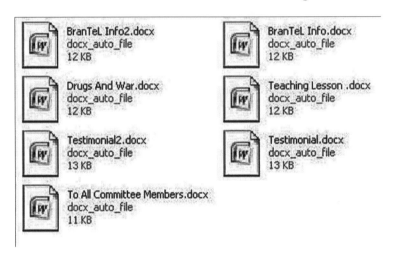

Figure 4-8

Right-click on the file To All Committee Members and select copy form the menu (see Figure 4-9)

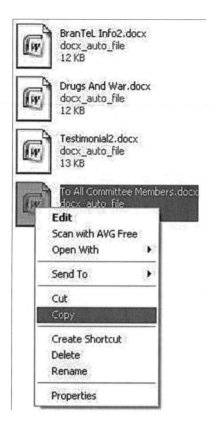

Figure 4-9

Right-click on the My Documents folder and choose Paste from the menu (see Figure 4-10)

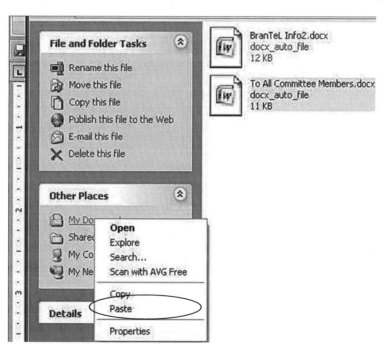

Figure 4-10

Now the file "To All Committee Members" is located in both folders (Word 2007 and My Documents). That is all there is to copying a file.

Copy all of the files that you downloaded to the Word 2007 folder.

Chapter Four   Review

Organizing your files and folders can make your life easier. Putting similar files together inside of a folder can make the files easier to find.

Clicking the File command on the Menu Bar will allow you to create a new folder. You should give the new folder a new name and the name should give you a clue as to its contents.

You can drag files from one folder to another folder. You can also right-click on a file and choose cut or copy from the shortcut menu.

Chapter Four   Quiz

Create a new folder inside the My Document folder.

Name the folder My Files.

Move two of the Word files to the new folder.

Chapter Five   Formatting Text

You have probably seen documents that use different fonts, italicized and boldfaced type, and fancy paragraph formatting. This chapter will explain how to format both characters and paragraphs. You will learn how to change the appearance, size, and color of the characters in your documents. You will also learn how to format paragraphs: aligning text to the left, right, and center of the page. You will also learn how to add borders to paragraphs and how to create bulleted and number lists.

Knowing how to format characters and paragraphs gives your documents more impact and makes them easier to read.

Lesson 5 – 1 Formatting Text using the Ribbon

Word allows you to put emphasis on text in your document by making the text darker and slightly heavier. This is called Bold. You can also make the text slanted (italics), or make the text larger (or smaller), or you can use a different typeface.

The easiest way to apply character changes is to use the Font Group of the Home tab of the Ribbon.

Open the document titled Agenda

This document is included with the files that you downloaded. We can use this document to explore some the different formatting options available.

Select the top line of the document

We will use the Font Group to make some simple formatting changes. This group is found on the Home tab of the Ribbon and is shown in Figure 5-1.

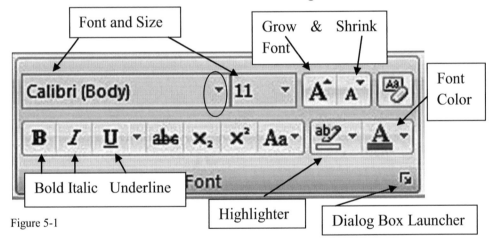

Figure 5-1

First we will change the typeface to something different than the default font.

Click the drop down arrow so we can change the font we are using (where the circle is in the figure)

Find Times New Roman **from the list and click on it**

The font for the top line will immediately change from Calibri to Times New Roman. This is better but we need it to stand out even more. To help accomplish this we can make the font larger.

Click the drop down arrow for the font size (the one next to the number 11)

Click on the number 20

Now the font is not only a different type than the rest of the document it is also larger.

Select the next two lines and change them to Times New Roman and a size of 14

It is starting to look better, but AGENDA still needs some work.

Click the mouse inside the word AGENDA

Click the icon for Italic

Agenda will change to *Agenda*. That is not quite what I wanted, so click the Italic button a second time to put AGENDA back the way it was.

With the insertion point still inside the word AGENDA **select the font** Monotype Corsiva

Now AGENDA stands out, but I still think it needs something.

With the insertion point still inside the word AGENDA **click the Grow Font button until the Font Size is 16**

That is much better, but it still needs something.

Note: Each time you click the Grow Font button the font size will increase by one point and each time you click the decrease font button the font will decrease by one point. Just so you know, fonts are measure in points and each point is 1/72 of an inch.

With the insertion point still inside the word AGENDA **click the underline button**

Now I am happy, it looks good. Say, what if we changed the color of the font to make it stand out even more?

With the insertion point still inside the word AGENDA **click the down arrow on the Font color button**

From here we can choose any of the theme colors or the standard colors. If the color you want is not shown you can click on the More Colors choice at the bottom.

Click on the Red choice from the Standard Colors

I thought that I was happy before but this is even better.

Save your changes

Lesson 5 – 2   Using the Dialog Box

There are a lot of things that you can do from the Ribbon, but there are even more formatting things that we can do. These need to be done from the Font Dialog Box.

If necessary open the Agenda **document**

Select the text Call to Order

Click the Font Dialog Box Launcher

The Font Dialog box will jump onto the screen and is shown in Figure 5-2.

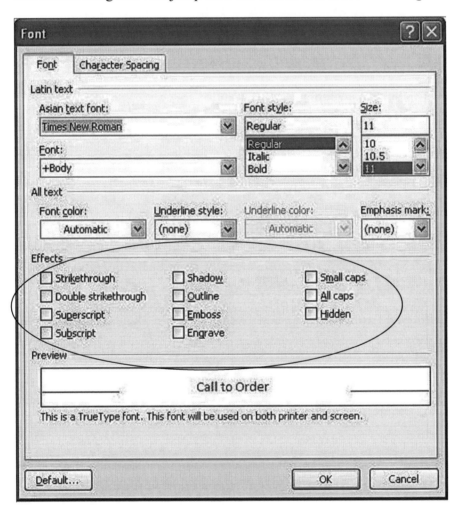

Figure 5-2

The top half of the dialog box has most of the items that are found on the Ribbon, but the bottom half has some new things on it. First we have the effects section and under that we have the preview area. The preview area has the selected text in it and allows us to see how any changes made will look in the document.

Each effect has a checkbox next to it. Clicking the effect checkbox once will turn the effect on and clicking it a second time will turn the effect off.

Using the mouse click each effect on and then off while watching the preview area

Note: If you make any changes and then click the Default button on the bottom left these changes will be the default settings for all new documents.

After you are finished experimenting click the Cancel button

Close the document without saving any changes

Lesson 5 – 3   Paragraph Alignment & Shading

If you are one of those people who like everything nice and neat and even, this lesson just might make your day. In this lesson we will be discussing how your text is aligned on the page. We will also discuss line spacing and background shading.

Open the document titled Advertisement

This document can be found with the files that you downloaded.

This document contains an advertisement that we were going to run in a local magazine. We have decided that we need to dress the document up a little before we submit it to the magazine.

Select the first line of the document

Using the techniques you have learned in the past two lessons change the font to Arial and the size to 18

Change the font color to Red Accent 2 (See Figure 5-3)

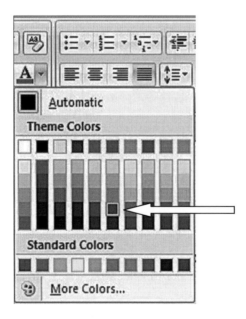

Figure 5-3

The first line is starting to get there, but it is not quite there yet. It would look a lot better if the text was centered in the page.

With the text still selected click the Center Align button in the Paragraph Group of the Home Tab (see Figure 5-4)

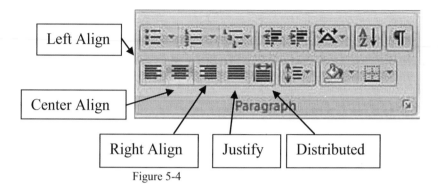

Figure 5-4

The text will now be centered on the page and will look much better.

Select the following words in the next line and make then bold

> Quality
> Dedicated
> hard working

In the next paragraph select excited and motivated **and make both words bold**

Select the third paragraph

> The first paragraph in this document is center aligned on the page. The second paragraph is aligned to the left. Now we will experiment with the third paragraph.

Click on the Justify button

> This will align the text on both the right and the left. Extra spaces will be added as necessary to accomplish this. The top line did not change very much and you may have to use the undo button and then justify it a second time to see the change.

> The Distributed button is similar to the Justify button in its description with an exception. With the Justify option the formatting is stopped when the text stops. With Distributed the formatting stops at the end of the line. The text is spread out to cover the entire line.

With the paragraph still selected, click the Distributed button

> The result of clicking the Distributed button is a lot different than clicking the Justify button.

Select the last paragraph and then click the Justify button

Both side of the text are nice and even and it looks like you spent a lot of time making sure everything was perfect.

While the paragraph is still selected, let's check the spacing of the lines in the paragraph.

Click the Line Spacing button just to the right of the Distributed button

A menu will drop down to allow you to select the spacing you would like to use. This is shown in Figure 5-5.

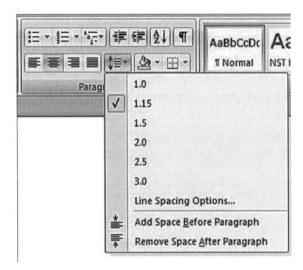

Figure 5-5

As you can see the default spacing between the lines is 1.15. You can change the spacing by simply clicking on the desired spacing.

Click on 2.0 and see the difference it makes

Note: You do not have to select the entire paragraph to make these changes. If the insertion point is anywhere inside the paragraph, it will act the same as if you selected all of the text in the paragraph.

Change the spacing back to 1.15

With the insertion point inside the paragraph, click the Add space before paragraph and carefully watch the spacing before the paragraph

You should notice the paragraph move slightly down and a small space being added to the document. If you missed it and want to try it a second time, you will first have to remove the space you added before you can add a space before the paragraph.

With the insertion point still inside the paragraph, remove the space after the paragraph

As you can see the last paragraph moved up toward the paragraph above it.

Add a space below the paragraph to return it to its original position

Before we finish this lesson there is one more thing I want to show you. We all know that if we have a printed document and we want something to stand out, we go over the text with a highlighter. Microsoft thought that it might be a good idea to have a feature that would allow us to insert a highlighted section into our document and everyone viewing or printing the document would already have that part highlighted.

In the second paragraph select the word quality

Click the down arrow on the Highlight Command

The Highlight Command is located in the Font Group of the Home Tab and is shown in Figure 5-6

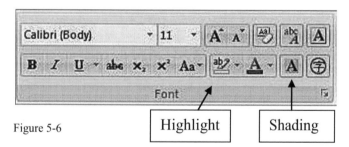

Figure 5-6   Highlight   Shading

Move the mouse over each color and watch the effects it has on the selected section

When you are finished choose the yellow color

Repeat this for the words: dedicated and hard working

Shading is similar to this, only the color is always a light gray.

Select the third paragraph and change the alignment to Justify and then click the Shading button

Save your changes and then close the document

Lesson 5 – 4   Numbered and Bulleted Lists

There may come a time when you want the reader of your document to follow specific instructions. It would be great if the instructions were in a numbered fashion so they could be easily followed. In this lesson we will create a numbered list. We will also create a list using bullets instead of numbers.

Create a new blank document

This can be done by using the Office button and choosing new then blank document, or you can click the new shortcut on the Quick Access toolbar.

Type the following and then press Enter

The proper way to turn on your computer

Click the down arrow on the Numbered List button

This is located in the Paragraph Group and is shown in Figure 5-7.

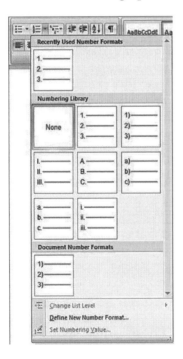

Figure 5-7

Note: If you had previously used the numbered list, there would be a section above the Numbering Library showing the previously used numbering lists.

Click on the first choice, the one with a number and then a dot

The number 1 appears on the page. Now you can type the first thing that you want in your list.

Type Take a deep breath **and then press Enter**

Immediately the number 2 will appear on the page ready for you to type the second item in your list.

Type Exhale **and then press Enter**

The number 3 will appear on the screen ready for you to type the third item in the list.

Type Press the On/Off button **and then press Enter**

The number 4 will appear on the screen. We don't have a forth point to our list so we need to get rid of this number.

Press the Enter button again

This will get rid of the number 4 and take the insertion point back to the left margin.

Save this document as Numbered List and then close it

If the order of the items in your list is not important, a bulleted list might work even better that a numbered list.

Create another new document

In this document we will create a bulleted list using a recipe for making Almond Cinnamon Balls.

Using the bulleted list will be great for listing the ingredients, since just listing the ingredients does not require doing something in a specific order.

Type Almond Cinnamon Balls ingredients (makes 15) **and center it on the page**

Press Enter twice

This will create a blank line between the title and the body

Change the font size for the title section to 16

This will make it stand out

Click the downward pointing arrow on the bulleted List

The Bullet Library dialog box will come to the screen as shown in Figure 5-8.

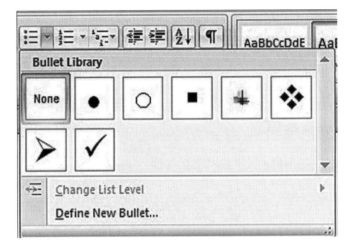

Figure 5-8

From here we can choose which bullet we want to use in our list. The bullet is the small dot or shape that we will have in front of every item in our list.

From the library choose the bullet with the four small diamonds by clicking on it

When you click on the desired bullet, the dialog box will disappear and the first bullet will appear in your document. The insertion point will be directly after the bullet.

Type 1 ½ Cups ground almonds **and the press Enter**

When you press the enter key the second bullet will appear, ready for you to type the next item in the list.

Add the following items to the list by typing each item and then pressing Enter

1/3 Cup granulated sugar

1 Tablespoon ground cinnamon

2 white eggs

Oil for greasing

Confectioner' sugar for dredging

When you have entered the last item and pressed the Enter key, there will be another bullet added and the insertion point will be to the right of it, just as there was every other time. Silly Word doesn't realize that you are finished with the list. Now we have to get rid of the last bullet.

Press the Enter key a second time to get rid of the last bullet

Save your document as Bulleted List and then close it

Note: If you start a paragraph with an asterisk or the number 1 then a period, Word will assume that you want to start a new bulleted or numbered list and start the process for you. If you do not want the numbered or bulleted list, click on the AutoCorrect options button. The AutoCorrect options button is shown in Figure 5-9 & 10. From here you can stop or undo the automatic numbering.

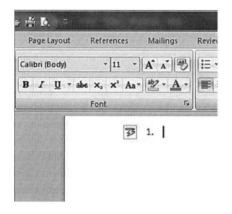

Figure 5-9

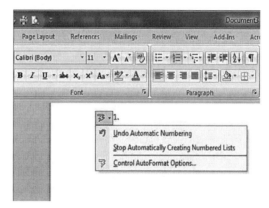

Figure 5-10

Lesson 5 -5   Adding Borders to your Paragraphs

Borders are lines that you can add to the top, bottom, and sides of your paragraphs. Borders can help paragraphs stand out from the rest of your document. In this lesson we will add a border to an existing document.

Open the Advertisement **document**

Click the mouse somewhere in the top line

Since the borders command is in the paragraph Group, we only have to be somewhere inside the paragraph to add a border to the entire paragraph.

Click the down arrow on the Border Command

A list of all the borders you can access will drop down onto the screen as shown in Figure 5-11.

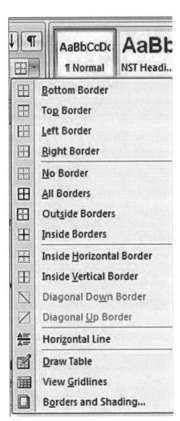

Figure 5-11

Click the bottom border option

A line will appear in the document under the first paragraph. Don't forget that even though it is only one line it is a paragraph.

Click inside the last paragraph

Add borders to the top, bottom, and both sides

For practice purposes I suggest that you add the borders one at a time, but you can add all of the borders at one time by selecting all borders from the drop down menu.

Save your changes and close the document

Lesson 5 – 6   Using the Format Painter

The Format Painter allows you to copy the formatting from one section of text and apply the same formatting to another section of text. This could save you a considerable amount of time if you are constantly applying the same formatting for different parts of your document. In this lesson we will use the Format Painter to copy the formatting from one section of text and apply it to another section of text.

Open the document titled First Quarter Review

This file can be found with the files that you downloaded.

Select the first line of the document and change the formatting to Bold, Italic, and Underlined

The document should now look like Figure 5-12.

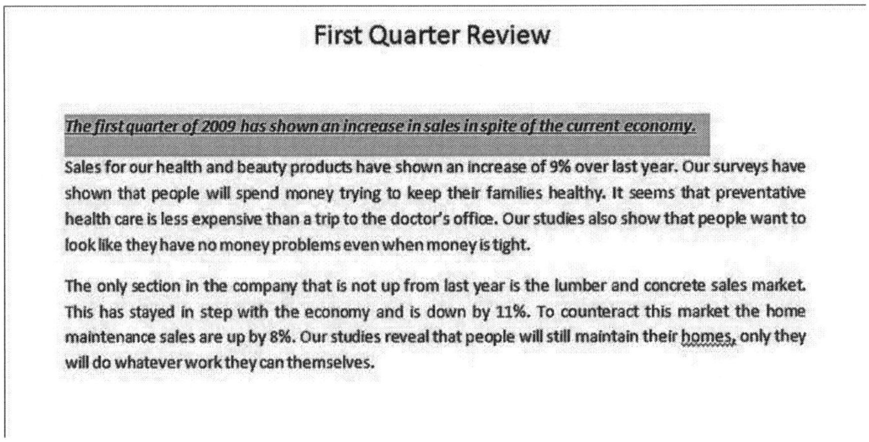

First Quarter Review

The first quarter of 2009 has shown an increase in sales in spite of the current economy.

Sales for our health and beauty products have shown an increase of 9% over last year. Our surveys have shown that people will spend money trying to keep their families healthy. It seems that preventative health care is less expensive than a trip to the doctor's office. Our studies also show that people want to look like they have no money problems even when money is tight.

The only section in the company that is not up from last year is the lumber and concrete sales market. This has stayed in step with the economy and is down by 11%. To counteract this market the home maintenance sales are up by 8%. Our studies reveal that people will still maintain their homes, only they will do whatever work they can themselves.

Figure 5-12

Now all we have to do is copy the formatting from here and paste it to another section of text.

With the text still selected click on the Format Painter

The Format Painter is located on the Home Tab in the Clipboard group (See Figure 5-13).

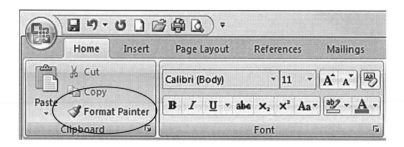

Figure 5-13

Before we can copy formatting, we have to let Word know what we are copying that is why the text must be selected before we click on the Format Pinter.

As you move the mouse into the text area the pointer will be replaced by a small paint brush. To paint the formatting to other text click the mouse at the beginning of the text you want formatted and drag the mouse to the end of the text you want formatted.

Move the mouse pointer to the first sentence of the second paragraph

Click the mouse and hold the left mouse button down

Drag the mouse to the end of the sentence

Release the left mouse button

Once you release the mouse button the painting will stop.

Repeat this process to paint the third sentence of the third paragraph

When you are finished the document should look like Figure 5-14.

First Quarter Review

The first quarter of 2009 has shown an increase in sales in spite of the current economy.

Sales for our health and beauty products have shown an increase of 9% over last year. Our surveys have shown that people will spend money trying to keep their families healthy. It seems that preventative health care is less expensive than a trip to the doctor's office. Our studies also show that people want to look like they have no money problems even when money is tight.

The only section in the company that is not up from last year is the lumber and concrete sales market. This has stayed in step with the economy and is down by 11%. _To counteract this market the home maintenance sales are up by 8%._ Our studies reveal that people will still maintain their homes; only they will do whatever work they can themselves.

Figure 5-14

Note: If you have several items (or more than one) you want to paint with the Format Painter, double click the Format Painter instead of single clicking on it. Double-clicking the Format Painter will keep it turned on until you decide to turn it off. It can be turned off by clicking the Format painter again or by pressing the ESC key on the keyboard.

Save your changes and close the file

Chapter Five   Review

Word allows you to put emphasis on text in your document by making the text darker and slightly heavier. This is called Bold. You can also make the text slanted (italics), or make the text larger (or smaller), or you can use a different typeface. You can also change the size of the font. If you desire, you can change the color of the text and add certain effects to the text.

Most formatting changes are done from the Font Group on the Home Tab. Other changes can be done from the Font Dialog box.

Paragraphs can be aligned on the right or left side of the page. Text can also be aligned with the center of the page. Text can also be aligned on the right and the left side by using the Justify command.

You can adjust the spacing between the lines of the paragraph as well as adding or removing the spacing between the paragraphs.

Text in the document can be highlighted just like you can highlight it on paper. Shading is similar to highlighting only the color is always light gray.

If you need steps to be followed in a specific order, use a numbered list otherwise use a bulleted list.

Borders can be added to your paragraphs to help them stand out from the rest of the document.

The Format painter will let you easily copy the formatting from the selected text to another section of text.

Chapter Five   Quiz

1) The Font Group is located on which tab of the Ribbon?
2) When you are adjusting the size of the font, the rule is: The larger the number, the larger the font. **True or False**
3) Explain the difference between the "All caps" effect and the "Small caps" effect.
4) If you click on the Justify command in the Paragraph Group, which edge(s) of the paragraph will be aligned?
5) If you click on the Line Spacing command in the Paragraph Group, you can either add or remove spaces before or after the paragraph. **True or False**
6) What type of list is preferred if you need steps to be performed in a specific order?
7) If you desire, you can have a bulleted list start with a check mark. **True or False**
8) If you wanted a line below a paragraph, which border you add to that paragraph?
9) When using the Format Painter, the first thing you do is click the Format Painter command on the Ribbon. **True or False**
10) How do you turn the Format Painter on and have it stay on until you decide to turn it off?

Chapter Six   Using Styles and Themes

We are going to venture into an area which can save you a considerable amount of time in formatting your documents. Even though this is a wonderful feature, not many people know about it or how to use it. This feature is called styles. A style is a way to format text so that all formatting is the same every time you use this style. Since this is not used by a lot of people, and therefore not very well understood, it is best to show you how to use it rather than try to explain it. You will also see how to apply a style to any text, and how to automatically update all instances of that style in a document. This is great because if you decide to make a change to the style, every time that style was used it will also be changed. You do not have to go back and reformat every instance of the old style. With styles you can choose one of the many predefined styles or you can create your own new style. In this chapter, we will make our own new style that we can apply to any document.

After we have discussed styles, we will discuss Themes and how themes differ from styles.

Lesson 6 – 1   Creating a New Style

Before we create a new style, we must consider that there are many pre-made styles available for our use. If we want to use a pre-made style we need to apply the style to the text. Lesson 6-2 will cover applying a style.

Open the file To All Committee Members2.docx

This file can be found in the files that you downloaded.

The first thing that we need to do is format some text. Once we are satisfied with the formatting we can save it as a style. After we save it as a style we can use it whenever we want.

Select the first line of the document and then click the Font Dialog Box Launcher

The Font Dialog box launcher is the small arrow on the bottom right of the Font Group which is on the Home Tab (shown in Figure 6-1).

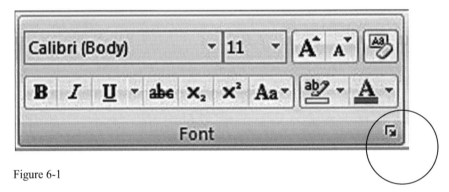

Figure 6-1

The Font Dialog Box will spring onto the screen with all of the formatting choices that are available to you. The dialog box can be seen in Figure 6-2.

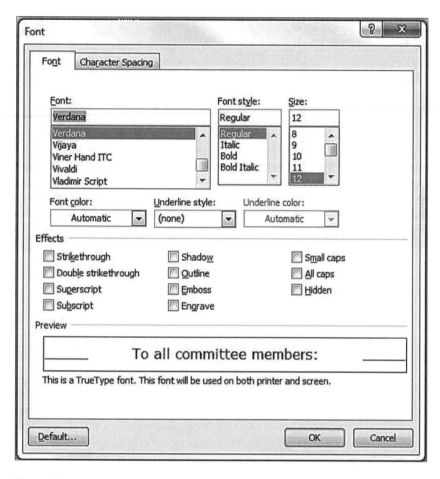

Figure 6-2

In the Font section, scroll down until you find Verdana and then click on it

This will change the font from the default setting to Verdana.

Click on Bold in the Font Style Section

This will change the text to Bold.

Scroll down to and click on the number 16 in the Size section

The font will now be larger than the rest of the text in the document.

Next we should change the color of the font to a red.

Click the down arrow under Font Color and then select more colors

This is shown in Figure 6-3.

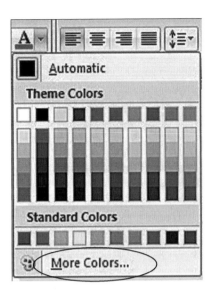

Figure 6-3

From the Standard colors choose the dark red at the bottom (See Figure 6-4)

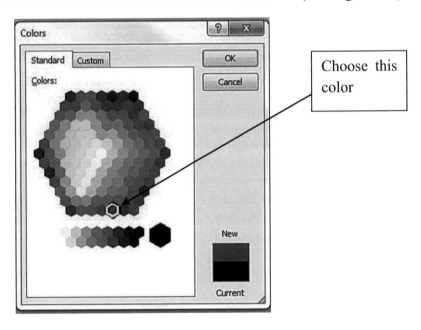

Figure 6-4

Click the OK button

Lastly click on the checkbox next to Small Caps then click OK

The last formatting change we will make for this section of text is to center the text on the page.

In the paragraph Group click on the Center Align button (See Figure 6-5)

Figure 6-5

The formatting is now complete and we want to save this as a style. Since there is not a pre-made style like this we will have to save this with a unique name. Giving this style a unique name will separate it from all other styles in the Styles Group.

Click the More Button in the Styles Group (See Figure 6-6)

More
Button

Figure 6-6

This will bring a new series of choices to the screen. We will want to save this style (the selected text) as a new quick style (See Figure 6-7).

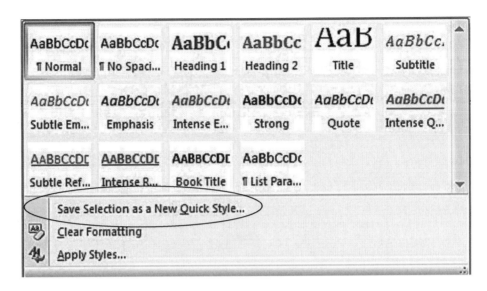

Figure 6-7

Click on the Save as a New Quick Style choice

The Create New Style from Formatting Dialog box will come onto the screen. All that is left for us to do is give the style a name. The dialog box is shown in Figure 6-8.

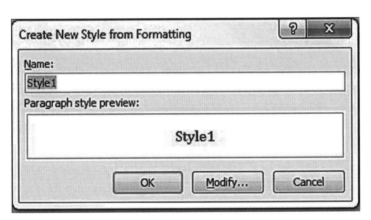

Figure 6-8

Under Name type Committee Heading **and then click OK**

This will save our style with the name Committee Heading and will be visible in the Styles Group as shown in Figure 6-9.

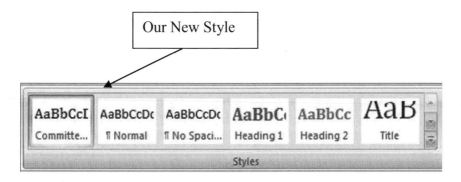

Figure 6-9

Now we have created a new style but it is only good for this one document. That doesn't do us much good does it? If you want this style to be available for all documents we need to do a little more work.

Right-click on the Committee Heading style we just created

Click on Modify

This will bring the Modify Style Dialog Box to the screen as shown in Figure 6-10.

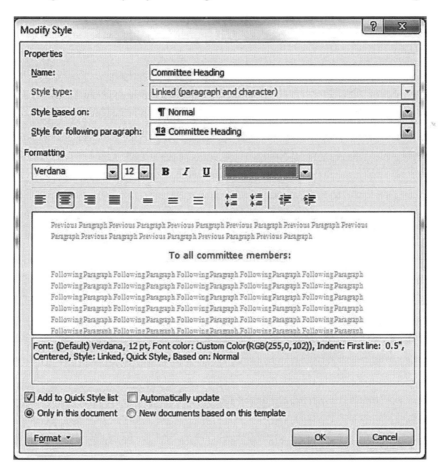

Figure 6-10

There is a lot to this dialog box, so let's try to go over the choices you can make. First we have the name of the style which we can keep or change. Let's keep the name the same. This style is going to be based on the Normal style, which is fine so we can leave this like it is. The next part is the style for the following paragraph. When we press Enter and start a new paragraph, what style do we want to start using? It would be very unusual for a heading to be followed by another heading, so we will want to change this from Committee Heading to Normal. It makes sense that after we have typed the heading for the paragraph the next paragraph should be normal text.

Click the down arrow for "Style for following paragraph" **and select Normal from the drop down list**

We can use the formatting, if we choose, to change the font, and size color of the text. We could also change the bold, italic, and underline choices.

If we wanted we could also change the alignment of the text and the line spacing.

Now we want to look down to the bottom of the dialog box. You will notice that the radio button next to only in this document is marked. If we want this style to be available for all new documents that we create, we have to click on the new documents based on this template choice.

Click on the radio button next to new documents based on this template

Just above this is another important choice. This choice is whether or not to automatically update. This may seem strange but here is how it works. If we make a change to this style we can have Word automatically go through the document and update every occurrence of the style in the document. Can you imagine how inconvenient it would be to have to go back into the document and manually change every occurrence of the style because we decided to change the style just a little bit?

Click the checkbox next to automatically update and then click OK

Now every time we start a new document this style will be available for us to use.

Save your changes and close the document

Lesson 6 – 2　　　Applying a Style

In lesson 6-1 we made a new style that we can use for any text in the document. To use the new style we need to apply the style to the selected text.

We have to remember that this is a paragraph style. That means if we click anywhere inside of a paragraph and then apply the style, the entire paragraph will use that style.

Open the To All Committee Members2 **document if it is not already open**

Click anywhere inside the paragraph that starts with "The July Meeting"

We are now inside the paragraph and we will now apply the style. Applying the style is a simple process. All we have to do is click on the style we want with the mouse.

Click on the Committee Headings style

The entire name Committee Headings will not be visible as is shown in Figure 6-11.

Figure 6-11

As soon as you click on the style, everything in the paragraph will change to reflect this style. A quick look at your document will show you that this is probable not what you wanted.

Click the Undo button on the Quick Access Toolbar

This will put the paragraph back to the original settings. This time we will be a little pickier about what we change.

Select the word July in the third paragraph

Selecting one word will allow us to change the formatting of just this word to the style we need. You use the same procedure for applying the style to the selected text as we did to apply the style to the entire paragraph. All we have to do is click on the style that we want to apply.

Click on the Committee Headings style

As long as we are looking at styles, it would only make sense to look at the pre-made styles and see if we like any of them.

Click inside the fourth paragraph, the one that has Bill Mills Committee chair in it

As long as we inside the paragraph we can see what a style looks like without actually applying the style. If we move the mouse pointer over one of the styles the text will change to match that style. When we move the mouse pointer to the next style the text will change to match that style.

Move your mouse pointer to the first style on the left and let it pause there for a few moments and then repeat this process for each of the styles that are visible

The text in the paragraph will change to match the style that the mouse pointer is resting upon. But wait! There are more styles that just these to choose from.

Click the More Styles button

If you have forgotten where this is located refer to figure 6-6.

Move the mouse over each of the available choices and watch the results

Styles may not seem like a big deal to you and not worth the time to mess with. If this is the case consider this book. There are six styles that I use over and over throughout this book. There is the normal text, there are chapter titles and lesson titles, there is the text when you are to perform a task, under each figure there is a different text style, and there is a different style for the text you are suppose to type.

Using styles can make you life a lot easier if you will only use them.

Close the document and save your changes

Lesson 6 – 3   Choosing a Theme

As we saw in the last two lessons a style will change the formatting of selected text or paragraph. A theme is a little different. A theme will affect the entire document.

Open the document titled To All Committee Members3 **from the downloaded files**

This document will allow us to see how the different themes will affect the document. The Themes Group is located on the far left side of the Page Layout Tab of the Ribbon. The Themes Group is shown in Figure 6-12.

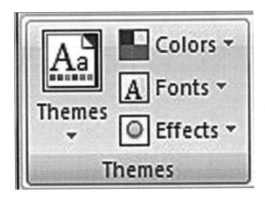

Figure 6-12

Click on Themes on the left side of the group

Using themes will let Word apply constant fonts throughout the document as well as font colors and typeface. Your document will look very professional with themes.

The available themes will drop down and you can choose the theme that best suits your needs. This is shown in Figure 6-13.

As you can see, the Office theme is applied to your document by default but you can choose to change this.

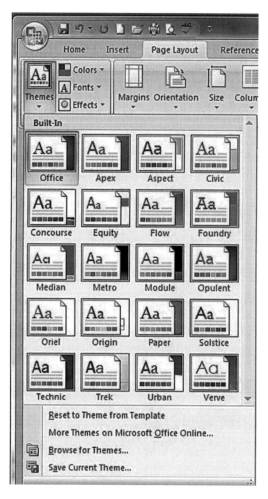

Figure 6-13

Just like with the styles you can move the mouse pointer to each theme and the document will change to match this theme. Remember that a theme will change the entire document, not just some selected text or paragraph.

Move the mouse pointer over each theme and let it stop on each one for a few seconds

Watch the document to see how each theme will change the entire document. You will undoubtedly notice that with some themes the font color also changes. The reason for this will be explained next.

Suppose we found the theme we like with the font and typeface, but we would like the colors to be a little different? Next we will see how to change the colors associated with our document.

Click on the Drop down arrow next to Colors

A list of color schemes will drop down and you can choose the color scheme that will work for you. The first four colors in the scheme represent the font colors that are available and the last four colors represent the colors for tables and charts. You can see the color choices in Figure 6-14.

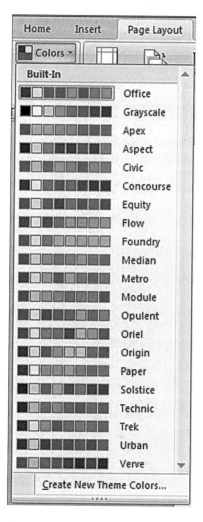

Figure 6-14

Move your mouse down through the list of choices and view each one as the text changes

When you find one that you like you can click on it to apply this color scheme.

This same procedure can be used to change the font for this theme.

If you want to keep this style of theme for use later, you can save it by clicking on Themes and choose Save at the bottom and give it a name.

Close the document without saving the changes

Chapter Six   Review

Styles will let you save and apply formatting to selected text. You can use the pre-made styles or format the selected text as you want and then save the formatting as a new style. When you save the style it can be only for the current document or for all new documents. If you decide to change the style, Word can automatically update all instances of the style to match the new style. To apply a style to other text, you need to first select the text you want to change and then click on the desired style.

Themes are similar to styles. While styles affect text and paragraphs, Themes affect the entire document. Themes affect the overall type of font used as well as the colors of the font, tables, and charts in the document. If you desire you can change the colors associated with a theme.

Chapter Six   Quiz

1) After you format the text, and you know that you will be using this formatting again, you will want to save it as a new _____.

2) Changing the text formatting can be done from the Font Group on the Ribbon or the Font Dialog Box. **True or False**

3) The command to right-align the text is found in which group of the Ribbon?

4) When you save a style, you must give it a unique name. **True or False**

5) What dialog box allows Word to automatically update all instances of a style?

6) New styles can only be used in the current document. **True or False**

7) If you want to apply constant fonts as well as font colors to the entire document you will use Themes. **True or False**

8) In the list of color schemes for the Themes the first 4 colors represent the _____ colors.

9) In the list of color schemes for the Themes the last 4 colors represent the colors for the _____ and _____.

10) Themes affect the formatting of text and styles affect the entire document. **True or False**

Chapter Seven   Working with Tables

Putting a table into your document will allow you to arrange information in a neat and organized grid. A table contains cells, where the text (data) is displayed, and the cells are arranged in columns and rows. You can format a table, such as giving it borders as well as shading and coloring options. As you will see, tables are very powerful but few people know how to use them, so they don't. In this chapter you will learn how to use tables.

Once you learn how to use tables, you might wonder how you have survived this long without using them. With a table you can:

Align text and numbers: Tables make it easy to align text and numbers in rows and columns.

Create a form: You can use tables to store lists names, addresses and phone numbers.

Track information: Word's mail merge feature actually stores information, such as names and addresses in a table. You can also copy and paste a table's information into a Microsoft Excel spreadsheet.

Create a publication: Tables allow you to create calendars, brochures, business cards, and many other publications.

Lesson 7 – 1   Creating a Table

In this lesson you will learn how to create a table and insert information into it. The first thing you need to do is insert a table into your document.

Create a new document by clicking the Office button and selecting New from the menu

Press Ctrl and Home to make sure your insertion point is at the top of the document

Press Enter once to leave a blank line at the top of the document

The blank line is not necessary, but it will make your life easier if you need to go back and insert something above the table, such as a title.

Click the down arrow in the Table group of the Insert tab of the Ribbon.

This will bring the Insert Table drop down List to the screen. This is shown in Figure 7-1.

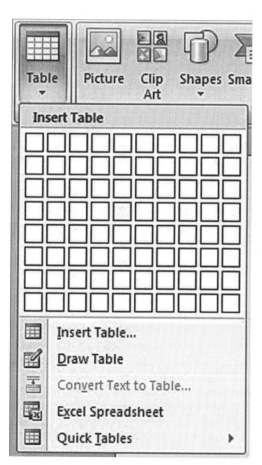

Figure 7-1

There are three ways you can insert a table. We will discuss the Insert Table method first.

Click on Insert Table

This will bring the Insert Table Dialog Box to the screen as shown in Figure 7-2

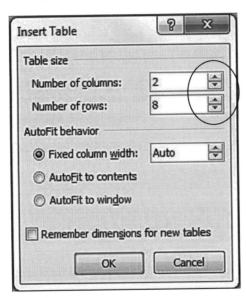

Figure 7-2

From here we can choose the number of columns and rows to include in our table. For our purposes we need 2 columns and 12 Rows. The fixed column width is fine for now; we will change the column width in a few minutes.

Click on the Up arrow in Rows until you have 12 showing

You may need to adjust the number of columns, if it is not at two

You may have to use the down arrow if the number of columns is larger than two.

Click OK to insert the table into your document.

Your document should look like Figure 7-3 with the table inserted.

Figure 7-3

I know this doesn't look like much yet, but it will in a few more minutes. First you need to know about cells and cell names. Each cell has a unique name that identifies its location in the table. The cell names start in the top left corner and go across and down as shown in Table 7-1.

A1	B1	C1	D1	E1	F1	G1	H1
A2	B2	C2	D2	E2	F2	G2	H2
A3	B3	C3	D3	E3	F3	G3	H3
A4	B4	C4	D4	E4	F4	G4	H4
A5	B5	C5	D5	E5	F5	G5	H5
A6	B6	C6	D6	E6	F6	G6	H6
A7	B7	C7	D7	E7	F7	G7	H7
A8	B8	C8	D8	E8	F8	G8	H8

Table 7-1

The columns are named A, B, C, D, E, etc. until you get to the end of your columns. If you have more that 26 columns, the next one after Z would be AA. The Rows are numbered 1, 2, 3, 4, 5, etc. until you run out of rows. The name of the first cell is A1.

169

This may seem silly to you, but as you will see later there is a reason for all of this madness. On the other hand, if you have worked with Microsoft Excel, a spreadsheet, this will seem like old hat to you.

The next thing we want to do is add some text into our table.

Click the mouse inside cell A1 to enter the first bit of text. When you have finished entering the text into the first cell, press TAB (not Enter) to move to the next cell.

Type the information as it appears in Table 7-2

Ctrl+B	Toggles the Bold formatting on and off
Ctrl+I	Toggles the Italicized font formatting on and off
Ctrl+U	Toggles the Underline font formatting on and off
Ctrl+Spacebar	Returns the font formatting to the default settings
Ctrl+O	Opens a document
Ctrl+S	Saves the current document
Ctrl+P	Sends the current document to the printer
Ctrl+C	Copies the selected text or object to the Windows Clipboard
Ctrl+X	Cuts the selected text or object from its original location and places it on the Windows Clipboard
Ctrl+V	Pastes any copied or cut text or object to the current location
Ctrl+Home	Moves the Insertion Point to the beginning of the document
Ctrl+End	Moves the Insertion point to the end of the document

Table 7-2

Well that is pretty good, but it would look even better if the line separating the two columns was more to the left side. The next step will show you how to adjust the column width.

Click on the line dividing column A from column B

The mouse pointer will change from a vertical line to a double line with a small arrow pointing to the right and to the left. Figure 7-4 shows a view of the double vertical lines.

Figure 7-4

When the mouse pointer changes into the two lines, you can click the left mouse button and drag the column width to a new location.

Click the left mouse button and hold it down. While you are still holding the left mouse button down drag the mouse to the left. As you drag the mouse to the left, a dotted vertical line will show you the current location of the line. When the line is just to the right of the word spacebar, release the mouse button.

The width of the column is now changed to match Figure 7-5.

Ctrl + B	Toggles the Bold formatting on and off
Ctrl + I	Toggles the Italicized font formatting on and off
Ctrl + U	Toggles the Underline font formatting on and off
Ctrl + Spacebar	Returns the font formatting to the default settings
Ctrl + O	Opens a document
Ctrl + S	Saves the current document
Ctrl + P	Sends the current document to the printer
Ctrl + C	Copies the selected text or object to the Windows Clipboard
Ctrl + X	Cuts the selected text or object from its original location and places it on the Windows Clipboard
Ctrl + V	Pastes any copied or cut text or object to the current location
Ctrl + Home	Moves the Insertion Point to the beginning of the document
Ctrl + End	Moves the Insertion point to the end of the document

Figure 7-5

The height of the rows can be changed using the same procedure only it will be positioned on the horizontal line separating the rows not the vertical line separating the columns.

Save this document as Table 1 and then close the document

Lesson 7 – 2   Storing Numbers & Text in a Table

Create a new blank document

This document will represent a typical document that might be found in a tour company. We will insert a table into this document, but we will use a different method that we used for the last table.

Change the paragraph alignment to center aligned and the font size to 18, and then type the word Tours and press Enter

Change the paragraph alignment back to left aligned and the font size back to 11

Click on Table on the Insert tab of the Ribbon

This time we will not use the Insert Table method to insert the table; there is an even easier way. Notice all of the small squares at the top. This represents the number of columns and rows of in the table. This table will have five columns and four rows.

Drag your mouse over the small squares until you have five cells across the top and four cells down highlighted.

As you drag the mouse over the squares, a table will be drawn in your document the same size as the selected cells.

When you have a table with five columns and four rows click the left mouse button.

A table with the desired number of columns and rows will now be inserted into your document.

Click the mouse inside cell A1 and type Destination

Press the Tab key on the keyboard

Pressing Tab will cause the next cell to the right to be the active cell. When the cell is active, you can type text or numbers into the cell (you could also insert objects into the cell, such as a picture).

In the active cell (B1) type Avg. Cost **and press Tab**

In cell C1 type Promotion **and press Tab**

In cell D1 type Projected Bookings **and press Tab**

In cell E1 type Projected Income **and press Tab**

Your table should now look like the table in Figure 7-6

Tours

Destination	Avg. Cost	Promotion	Projected Bookings	Projected Income

Figure 7-6

Using the information in Figure 7-7, enter the data into your table.

Tours

Destination	Avg. Cost	Promotion	Projected Bookings	Projected Income
Ottawa	$1,500	Yes	105	
Nova Scotia	$1,350	Yes	60	
Vancouver	$ 1,600	No	90	

Figure 7-7

Save this lesson as Tours Table

Lesson 7 – 3   Formatting a Table

In this lesson we will learn how to make the table look more attractive by formatting the table.

If it is not already open, open the Tours Table **document**

The first thing we should do is adjust the column width so that the Projected Bookings column is wide enough to hold the text on one line. While we are at it, let's adjust some of the other column widths that may not need to be as wide as they are.

Using the technique you learned in lesson 7-1, adjust the width of the columns until they are similar to the column widths in Figure 7-8

Tours

Destination	Avg. Cost	Promotion	Projected Bookings	Projected Income
Ottawa	$1,500	Yes	105	
Nova Scotia	$1,350	Yes	60	
Vancouver	$ 1,600	No	90	

Figure 7-8

The next thing we want to do is change the font style to bold and also center the text in the column. We only want to do this for the first row (the headings for the columns).

Move the mouse pointer to the left margin next to the row containing Destination, Avg. Cost, etc and click the left mouse button.

The first row of the table will now be highlighted and ready for us to format the text.

Click on the button with the "B" on it on the Home tab in the Font group.

This will cause the text in that row to be in the bold faceplate.

In the Paragraph group, click on the Center Align button

This will cause the text to be aligned in the center of each cell.

Now let's align some of the other text to the center of the cells.

Move the mouse until it is just above the word Promotion and touching the edge of the border of the column.

The pointer will change from the line that looks like a capital I to a dark filled-in arrow.

Click the left mouse button

The entire column will be highlighted. Now we can center the text in this column.

Click the Center Align button in the Paragraph group.

The text in the Promotions column will now be centered in the cells.

Repeat this procedure with the projected bookings column

There is something else you might want to consider. If you would like the table to be printed without the borders showing, follow the next steps.

Select the entire table by clicking in the left margin next to the first row and drag the mouse to the bottom of the table

The entire table will be highlighted. Now we can remove the cell borders form the table.

On the Home tab, in the paragraph group, click the down arrow on the Borders button

The drop down list will appear as shown in Figure 7-9.

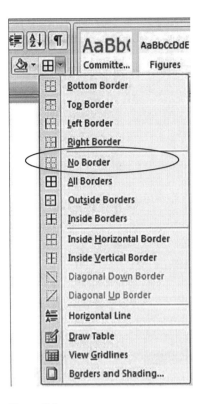

Figure 7-9

Select No Border from the drop down list

The table might show the borders in a dashed line format on the screen, but will not print the borders on the paper. Figure 7-10 shows how the document will look with no borders.

Tours

Destination	Avg. Cost	Promotion	Projected Bookings	Projected Income
Ottawa	$1,500	Yes	105	
Nova Scotia	$1,350	Yes	60	
Vancouver	$ 1,600	No	90	

Figure 7-10

If you want the borders to show in the table, repeat the above steps, only choose **All Borders** from the drop down list.

Save the document when you are finished

Lesson 7 – 4   Doing Mathematical Calculations

in a Table

A table is great for separating text into columns and rows, but this is not the true power of tables. The real power of tables is in doing mathematical calculations. A table is **not** an Excel spreadsheet; however it can perform simple calculations. Microsoft Excel is a computer program specifically designed to perform simple and complicated mathematical calculations. A table does not go that far, but we can do addition, subtraction, multiplication, and division quite easily using the built-in Formula button on the Data group of the Table Layout tab of the Ribbon. Wow that was a mouth full!

Do you remember, way back toward the beginning when I was talking about the Ribbon, I stated that some tabs were not always visible? When you insert a table into the document two new tabs are visible. One tab is the Table Design and the other is the Table Layout. These tabs will only be visible when the insertion point is inside the table. When the insertion point is outside the table, these tabs will again disappear.

Open the Tours Table **document if it is not already open**

Click the mouse inside the cell directly under Projected Income

The two new tabs will become visible on the Ribbon.

Click on the Layout tab

The Layout tab is shown in Figure 7-11. As you can see there are many additional feats that we can now perform. For this lesson, we will be using the Formula button on the far right side of the Layout tab in the Data Group.

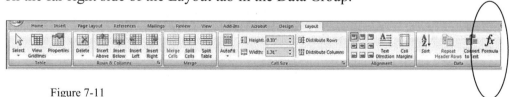

Figure 7-11

Click on the Formula button (the button with fx on it)

When you click on the Formula button, the Formula Dialog Box comes to the screen. This is shown in Figure 7-12. As you can see, a formula is already inserted into the Formula section of the dialog box.

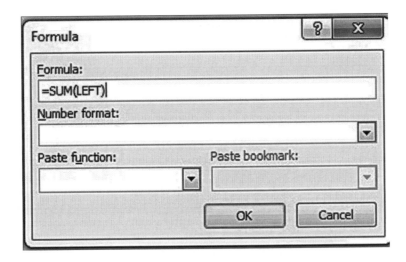

Figure 7-12

There are a few things you need to understand about formulas. The first thing is that all formulas must start with an equals sign. There are also some keywords. One of the keyword is shown in the dialog box, it is Sum. A sum refers to "addition" (adding numbers together). Another keyword that is shown is the word Left inside the parentheses. This is great, but it has some limitations: such as it will stop adding when it comes to a non numeric value. In our table, if we clicked OK, Word would start adding with the first column to the left (105) but it would stop when it came to the word Yes. The result is that Word would insert 105 into cell E2.

In our case we don't even want to add any numbers. By looking at the table you can see that to get the projected income, we would have to multiply the projected booking by the cost.

The first thing to do, is highlight everything inside the formula section then

Type the following in the formula section

=B2*D2

Then Click OK

Inside cell E2 the text should change to the following $157,500.00

Now, let me explain what you just did.

You started with an equals sign: all formulas must start with an equals sign.
You put in the first cell name: remember we talked about cell names and why you needed them, well this is why.
You put in the symbol for multiplication: the asterisk is the symbol for multiplication.
You put in the name of the second cell needed in the formula.
The resulting product of the multiplication is inserted into cell E2.

In English the formula would read: 1500 x 105 = 157,500

B2 * D2

Below is a table showing the mathematical symbols that you can use in a table.

Operator or Function Name	Purpose	Example
=	All formulas must start with an equals sign	=
+	Performs addition between values	=A1+B1
-	Performs subtraction between values	=A1-B1
*	Performs multiplication between values	=B1*2
/	Performs division between values	=A2/C3
SUM	Add all numbers in a range of fields	=SUM(ABOVE)
AVERAGE	Calculates the average of all the numbers in a range of fields	=AVERAGE(A2,B1, C3)
COUNT	Counts the number of items in a list	=COUNT(A2:C3)

Table 7-3

Using the just learned process, calculate the remaining two cells in the table

Save your work

179

Lesson 7 – 5   Adding Rows to a Table

If it is not open: open the Tours Table **document**

The Tours Table document should look like the document below.

Tours

Tours

Destination	Avg. Cost	Promotion	Projected Bookings	Projected Income
Ottawa	$1,500	Yes	105	$157,500.00
Nova Scotia	$1,350	Yes	60	$81,000.00
Vancouver	$1,600	No	90	$144,000.00

Figure 7-13

What happens if we need to add other items to the table, do we have to start all over again with a new table? Somehow I don't think Microsoft is going to do that to us. Let's add an additional row to the table.

Click the mouse in the last cell of the table (E4) and press the Tab button

A new row appears under the row 4, ready for us to input more data.

Add the following text into the cells (use the Tab key to move between the cells

Winnipeg	$1,200	No	50	

Figure 7-14

Finish the calculation for the Projected Income as you did in Lesson 7-4

The finished table should look like Figure 7-15.

180

Tours				
Destination	Avg. Cost	Promotion	Projected Bookings	Projected Income
Ottawa	$1,500	Yes	105	$157,500.00
Nova Scotia	$1,350	Yes	60	$81,000.00
Vancouver	$ 1,600	No	90	$144,000.00
Winnipeg	$ 1,200	No	50	$60,000.00

Figure 7-15

Adding an additional row to your table was easy, all you had to do is press the Tab key when you were in the last cell. What if you had to add a row in the middle of the document? Do we start from scratch all over again? Let's hope not. Now we will add a row between Nova Scotia and Vancouver.

Click your mouse so that the insertion point is in the cell with Vancouver in it

When the insertion point is inside a cell, we can insert, or delete, columns and rows. For our example we are going to insert a row above the row with Vancouver in it.

On the Layout tab and under the group Rows and Columns, click on the Insert Above button.

The Rows and Columns section of the Ribbon is shown in Figure 7-16.

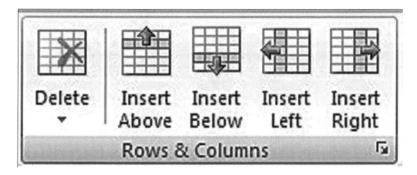

Figure 7-16

When you click on the button a new row will be inserted into the table directly above the row where the insertion point is. Your table should look like Figure 7-17.

Tours

Destination	Avg. Cost	Promotion	Projected Bookings	Projected Income
Ottawa	$1,500	Yes	105	$157,500.00
Nova Scotia	$1,350	Yes	60	$81,000.00
Vancouver	$ 1,600	No	90	$144,000.00
Winnipeg	$ 1,200	No	50	$60,000.00

Figure 7-17

Using the information in Figure 7-18 enter the information needed to make your table look like Figure 7-18 (use the formula for the Projected Income)

Tours

Destination	Avg. Cost	Promotion	Projected Bookings	Projected Income
Ottawa	$ 1,500	Yes	105	$157,500.00
Nova Scotia	$ 1,350	Yes	60	$81,000.00
Toronto	$ 1,050	No	65	$68,250.00
Vancouver	$ 1,600	No	90	$144,000.00
Winnipeg	$ 1,200	No	50	$60,000.00

Figure 7-18

You might want to take note of this: the formulas stayed with the correct column when you inserted the new row. Word knew that the cells, which were referenced in the formula, would be different and have new cell names. Word automatically changed the cell names in the formula to keep everything straight. Pretty cool, isn't it?

Save your work

Lesson 7 – 6   Merging cells in a Table

In a table you can merge cells together and put the data into one cell. Okay, why would you want to ever do this? How about under these conditions: You want to make a heading across the top of the table and you want everything in one nice clean cell.

In this lesson we will merge the top row of cells together into one cell and not have any of those cell borders to mess with.

If necessary, open the Tours Table **document**

Click the mouse just before the word Destination and drag it to the end of the word Projected Income

On the Layout tab of the Merge group, click on the Merge Cells button

This button is shown in Figure 7-19.

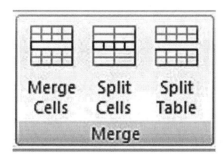

Figure 7-19

When you click on this button, you will get the results shown in either Figure 7-20A or 7-20B. I can hear you now, "What have you done to me?" Be patient, this will look a lot better in a moment.

Destination Avg. Cost Promotion Projected bookings Projected Income				
Ottawa	$1,500	Yes	105	$157,500.00
Nova Scotia	$1,350	Yes	60	$81,000.00
Toronto	$1.050	No	65	$68.250
Vancouver	$1,600	No	90	$144,000.00
Winnipeg	$1,200	No	50	$60,000.00

Figure 7-20 A

183

Tours

Destination Avg. Cost Promotion Projected Bookings Projected Income				
Ottawa	$ 1,500	Yes	105	$157,500.00
Nova Scotia	$ 1,350	Yes	60	$81,000.00
Toronto	$ 1,050	No	65	$68,250.00
Vancouver	$ 1,600	No	90	$144,000.00
Winnipeg	$ 1,200	No	50	$60,000.00

Figure 7-20 B

Click the mouse immediately after the word Destination and press the Delete button on the keyboard. Now press the spacebar several times until the Avg. Cost is directly above its associated column.

Now move the mouse to the end of Avg. Cost and press the Delete button and then use the spacebar until Promotion is lined up directly above the column with the Yes/No in it.

Now repeat the procedure and line up Projected Booking and Projected Income above their columns.

If your screen looks like Figure 7-20B, you may have to use the backspace before the word Destination to bring it to the right edge of the row and then adjust the spacing between the words.

The finished product should look like Figure 7-21.

Destination	Avg. Cost	Promotion	Projected bookings	Projected Income
Ottawa	$1,500	Yes	105	$157,500.00
Nova Scotia	$1,350	Yes	60	$81,000.00
Toronto	$1.050	No	65	$68.250
Vancouver	$1,600	No	90	$144,000.00
Winnipeg	$1,200	No	50	$60,000.00

Figure 7-21

Save your work when you are finished

Lesson 7 – 7   Adding Shading to a Table

Sometimes you want cells to stand out from the rest of the table. In our example of the Tours Table, we may want the heading row to stand out from the rest of the document. We can make it stand out from everything else by changing the background color of the cell. This is called shading. In this lesson, we will shade the top row of the table making it a different color.

Select the first row of the table

This can be done by clicking the left mouse button before the word Destination and dragging it to the end of the word Projected Listings. It can also be selected by clicking the mouse in the left margin next to the top row.

Click the Shading button that is in the Table Styles group of the Design tab

The Design tab is shown in Figure 7-22.

Figure 7-22

The Shading Color Dialog Box comes to the screen as shown in Figure 7-23

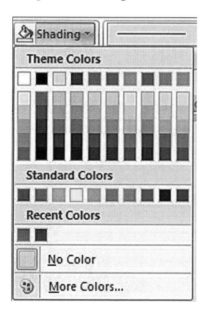

Figure 7-23

185

Move your mouse over the various colors to see what the selected row will look like using the various colors

When you have finished, choose the color orange

Save your work when you are finished

Close the Tours Table document

Chapter Seven   Review

Tables allow you to arrange information in a neat and organized grid. You insert a table from the Table Group on the Insert Tab of the Ribbon. You get to determine how many columns and rows are in the table.

Each cell in a table has a name that is determined by the column and row that holds the cell. Columns are labeled with letters starting with "A" on the left. Rows are numbered starting with 1 at the top. The very first cell is named "A1" (the letter always comes first in the name).

Additional rows can be added to the end of the table by clicking in the last cell and pressing the Tab key on the keyboard. If you need to add a row in a place other than the bottom, you need to click the mouse inside a cell either above or below where you would like the new row. Next you click on the Insert above or Insert below command that is in the Rows and Columns Group of the Layout Tab. From this group you can also insert a column to the left or to the right of the selected cell.

You can store text in a cell or you can store numbers in a cell. If you are storing numbers in the cell, you can use them to do mathematical calculations. This is done by using formulas. The Formula command is in the Data Group of the Layout Tab of the Ribbon.

All formulas must start with an equals sign. The data inside the cell is referenced by using the cell name.

Cells inside a table can be merged together to form a single larger cell. This is usually done when you need to make a heading row in the table.

Chapter Seven   Quiz

1) If you use the Insert Table command from the drop down menu, the number of rows and columns in the table will be preset and cannot be changed. **True or False**

2) In a table, what is the name of the cell that is located in row 2 and the third column from the left?

3) What happens if you hold the Ctrl key down and press the C key and then release both keys?

4) The width of a column, in a table, can be adjusted but the height of a row cannot be adjusted. **True or False**

5) If you want the text in a cell to show as italic, you would select the text and then click on which command in the Font group?

6) On which tab and group is the command to center align the text in a cell?

7) All formulas must start with what?

8) Write the formula to add the numbers in cells A3 and B4.

9) On what tab and group would you find the command to insert a row above cell D4?

10) Shading is when you change the _____ _____ of a cell.

Chapter Eight   Working with Graphics

So far we have done a lot of formatting and learning about the Ribbon. We will continue to use the Ribbon in almost every document we make. Right now let's look at another way to spruce up our documents. One thing that will make out documents look like a professional was hired to design them is inserting graphics into the document. In this chapter, we will try to cover most of the things a person can do with graphics. This chapter will have you design a series of pages for an annual meeting of the Branson Homeowners Association. These pages will detail some new properties we have purchased across the United States.

Lesson 8 – 1   Inserting Pictures

Create a new blank document

This can be done by clicking on the Office Button and selecting new.

Type the following:

Branson Homeowners Association ◄——— Press Enter Here

Goes Across the U.S.A.

Select the text and then change the font to Times New Roman

Change the font size to 36

Add the effect for small caps

Center align the text

These changes can be made by launching the font dialog box (except the center align which must be done from the paragraph group). To launch the dialog box, click the small button on the right side on the font group of the home tab. Figure 8-1 shows the location of the dialog box launcher.

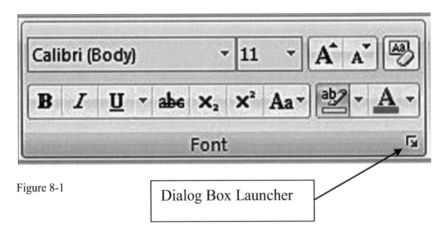

Figure 8-1   Dialog Box Launcher

The next thing we want to do is change the orientation from portrait to landscape.

Click on the Page Layout tab of the Ribbon

The orientation can be changed by using the Page Setup group and clicking on Orientation.

Click on Orientation

When you click on Orientation, a menu will drop down with two choices on it. You can either pick Portrait or Landscape. With Portrait the long side of the paper will go up and down. With Landscape the long side of the paper will be across the top and the bottom. This is shown in Figure 8-2.

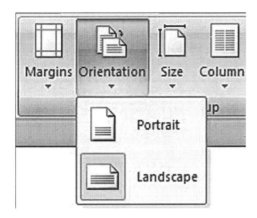

Figure 8-2

Click on the Landscape choice

The last thing we want to do for this page is to give it a background picture. We want the background to be almost see-through. For this effect we need to use a watermark. Word 2007 has several built-in watermarks available for us to use. For our document the built-in watermarks will not work, we will want to design our own.

On the Page Layout tab, click on Watermark in the Page Background Group

From here we could choose one of the standard watermarks, such as confidential or sample, but that would not help or document one bit. For our document we want something happier and a little more colorful. What we want is a custom watermark.

At the bottom of the choices, choose Custom Watermark

Clicking Custom Watermark will cause the Custom Watermark Dialog Box to come to the screen. This is shown if Figure 8-3.

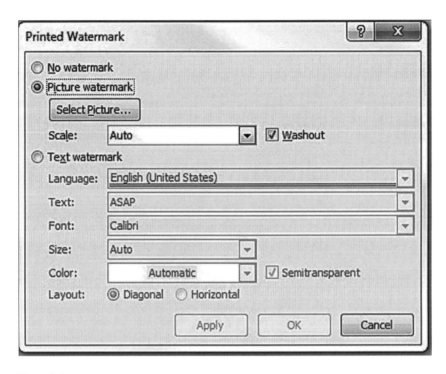

Figure 8-3

As you can see from this, we could choose not to have a watermark. We could also choose to use an existing picture as a watermark. Another choice is to use whatever text we want as a watermark. If we choose the text watermark, we can also choose the font and the size and the color of the text. In the text box, we can use one of the existing texts or we can type whatever text we want to appear in the watermark.

For our purposes we will use the Picture Watermark. This is already chosen by default.

Click on the Select Picture button

The Insert Image dialog box will come to the screen. From here we need to navigate to where the picture we want to use is located. The picture we want to use is located with the files that you downloaded.

Navigate to the folder where the downloaded files are located

Select the picture of the sun and then click the Insert button

This will put the path to the picture into the dialog box. The last choice we need to make is weather we want the background image to be washed out or not.

Make sure the check box next to Washed out is clicked and then click OK

This will put the image behind our text and provide a background for our first page. This watermark will be on every page of our document, so we want it to be almost invisible.

Click the mouse immediately before the word Branson and press Enter twice

This should center the text in the middle of the page

Save the document as Homeowners Association **and the formatting as a style called Homeowners**

The document should look like Figure 8-4.

Figure 8-4

Lesson 8 – 2   Working with Text Boxes

In this lesson we will continue with our series of pages for the Branson Homeowners Association. We will insert a US map and a text box indicating where Branson Missouri is located.

If necessary open Homeowners Association

Press Ctrl and End to get to the bottom of the document

Once we are at the end of the document we will need to add another page for our next insertion.

Click on the Insert tab of the Ribbon and choose Blank Page in the center of the Pages group

This will add a blank page just after the insertion point. The new page will also support our watermark. This is why we wanted to make sure the watermark was washed out. If it was not washed out, the image of the sun would be too dark and distract from our document.

The first thing we will want to do is check the orientation of the page and make sure it is Landscape

When you click on Page Layout and then Orientation, you will notice that Landscape is highlighted. This is because when you insert a blank page it follows the same orientation that is on the currently active page.

On the second page, type the following:

Branson Homeowners Association

Located in Branson, Missouri

Select the two lines and apply the Homeowners style to them by clicking on the Homeowners Style button in the Quick Style List

When you click on the style selection, the text will jump to the center of the page and the formatting will be the same as on the previous page.

Now we still need to insert a map of the United States so we can show everyone where Branson Missouri is located.

Click the mouse below the highlighted text and press Enter

This will put a blank line between the text and the picture of the map.

There is a picture of a U.S. map in the downloaded files. We will use that picture to mark the location of Branson Missouri.

Click the Insert tab and choose picture from the Illustrations group

This will bring the Insert picture Dialog box to the screen as it did in the previous lesson.

Navigate to and Click on the US Map and then click the Insert button

The picture of the US Map will be inserted into our document. Now all we have to do is mark where Branson is located and this page should be done.

On the Insert tab of the Ribbon, click on Text Box in the Text group

This will bring a list of the predefined text boxes on to the screen. We do not want to use a pre-defined text box, so we will draw our own box.

Click on Draw Text Box down by the bottom

The mouse pointer will change into a plus sign (+).

Move the mouse until the plus sign is somewhere around Pennsylvania and click and hold the left mouse button down and drag the mouse to the right and down until the box is about the same size as the one in Figure 8-5.

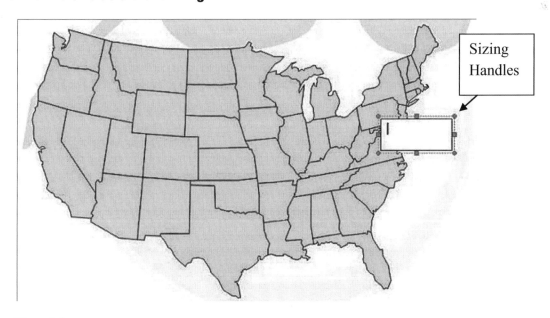

Figure 8-5

You will notice that the insertion point is inside the text box waiting for you to type something in it. What good is text box with no text?

Type Branson Missouri **inside the text box**

It will look better if Branson and Missouri were on separate lines. If your text box is too small, the words will not fit inside the box. If the box is too large, the text may only be on one line. We will use the sizing handles on either the bottom and/or the left side to make the box the correct size.

If the height of your text box is to short (or to tall), position your mouse pointer over the small square in the center of the bottom line of the text box and click the left mouse button and drag the mouse downward (or upward) until the box is large enough for two lines of text.

If the width of your text box is to long (or to short), position your mouse pointer over the small square in the center of the right line of the text box and click the left mouse button and drag the mouse to the right (or to the left) until the box is large enough for one word.

Your text box should look similar to Figure 8-6.

Figure 8-6

Guess what, that is not where Branson Missouri is located! Let's put an arrow in our document from the text box to where Branson is on the map.

On the Inset tab, click on Shapes

A chart comes down showing the available shapes. This is shown in Figure 8-7.

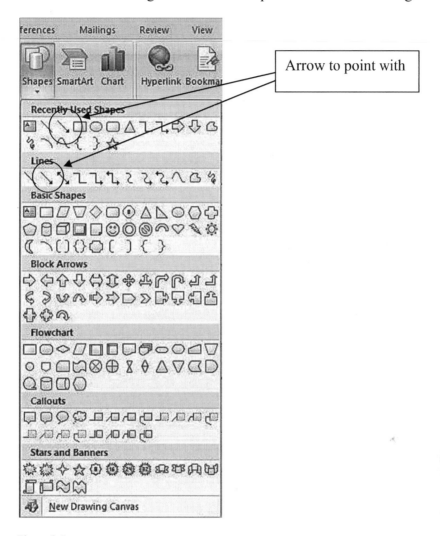

Figure 8-7

Click on the line with the arrow on the end of it

If we use this to point to the location of Branson, everyone will be sure to know where it is located on the map. When you click on the "arrow" the mouse pointer will change into a plus sign just as it did when we inserted the text box.

Move the mouse pointer until it is touching the left edge if the text box

When it is touching the edge of the text box, click and hold down the left mouse button. Using your mouse draw the line to the Southwest part of Missouri.

Your drawing should look similar to Figure 8-8. If you need to reposition your line, simply click on the line with the mouse and a dot will appear at each end of the line. When the dots appear, you can click on either dot and re-drag the line to a new location. If you would like you could just hit the delete button after the dots appear and repeat the above steps to draw a new line.

Figure 8-8

Save your work

Lesson 8 – 3   Working with Clipart

You might have noticed that some documents have interesting pictures, almost works of art. Windows has a section called Clip Art that has some great pictures for us to use. In this lesson, we will insert some clip art into our document.

Open the document called Homeowners Association **and press Ctrl and End to get to the end of the document**

The end of the document is at the right side of the last picture we inserted. From here we will add another page to our document.

Press Enter on the keyboard

This should add a new page to our document. If it does not, press the Enter key until it adds a new page.

Type the following: Fun Times in Branson **on the first line of the new page and then press Enter**

It looks a little out of place just sitting there all by itself. It would probably look better if it looked like the other headings.

Change the style to match the Homeowners Style we made earlier

You can do this by clicking the mouse anywhere on the newly added words and clicking on the Homeowners button in the quick styles. Now it looks like the rest of the headings.

Now that the heading looks like the other headings, we need to add some clip art showing how much fun you can have in Branson.

On the Insert tab, click on the Clip Art button

This will bring the Clip Art pane to the right side of your computer screen. This is where you can search through all of the Clip Art to find the one that will suit your purposes. By the way you can search online through a couple of hundred thousand clip art images if you can't find one that you like in the built-in clips. We will search for an image that will show things we can do in Branson.

Click your mouse in the search box at the top of the Clip Art pane

Figure 8-9 shows the clip art pane.

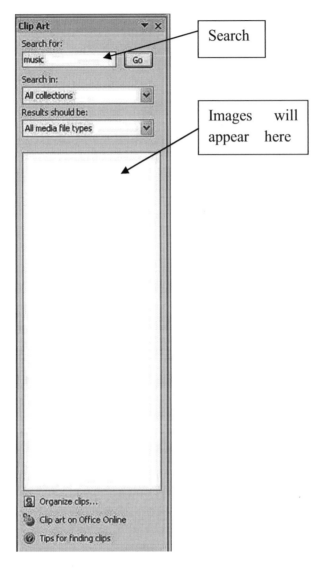

Figure 8-9

Type golf and then click on the Go button

In a few moments several small images should appear in the blank space. We should be able to find a golf picture we like.

Scroll down until you find the image of a female golfer shown in Figure 8-10

Click on the image

This will insert the image into our document.

Use this picture

Figure 8-10

Your document should look like Figure 8-11.

FUN TIMES IN BRANSON

Figure 8-11

This would be okay, if the picture was centered.

Click on the picture with your mouse and then click on the center alignment button in the paragraph section of the Home tab

This will center the picture between the left and right margins. Now, if it was only a little larger.

With the picture still selected, move your mouse to the lower right corner until your pointer changes into a double sided arrow at an angle.

Figure 8-12 shows the sizing handles for the picture. The sizing handles are where you can click and drag the mouse to resize the picture.

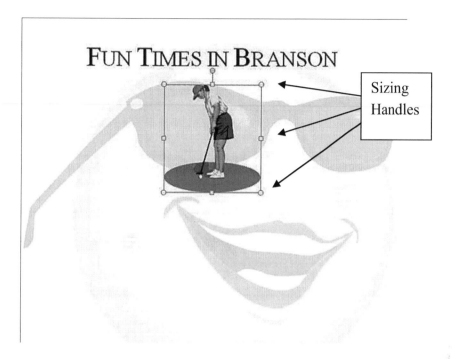

Figure 8-12

Click and hold the left mouse button and drag the sizing handle down and to the right until the picture almost fills the entire page then release the mouse button.

The finished page should look similar to Figure 8-13.

Save your work when you are finished

Figure 8-13

203

Lesson 8 – 4   Positioning Pictures

Have you ever wondered how some people have positioned their picture at strange angles? Not all pictures in your documents have to be aligned straight up and down. Pictures can be tilted to give your documents a real professional look. In this lesson we will practice tilting the pictures to different angles.

Open the Homeowners Association **document and press Ctrl and End to get to the end of the document**

Press the Enter key until a new page is added to the document

On the new page type:

> Branson Homeowners Association
>
> Our Newest Venture

Press Enter after each line

Select the two lines and change the style to the Homeowners style

We now have the headings and we will add another picture. The next picture we will add will be a picture of golf course.

Click the Insert tab and select Picture

Navigate to the Word 2007 folder (or the folder with the downloaded files) and select the picture Ireland_Golf

The picture of the golf course in Ireland will be inserted into the document, as shown in Figure 8-14.

Figure 8-14

Select the picture and center is as you did in the last lesson

This is accomplished by clicking on the picture and then clicking on the center align button in the paragraph group of the home tab.

As long as the picture is selected, the sizing handles are visible. There is also another handle that is visible. This handle is at the very top and is called the Free Rotate Handle. When you click on this handle and move the mouse to the right or left, the picture will rotate.

Click on the Free Rotate Handle and rotate the picture slightly to the left as shown in Figure 8-15

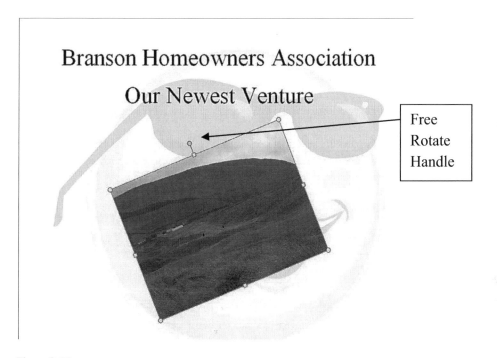

Figure 8-15

Pictures can be rotated to achieve the effect you need for your application.

Note: This photo was used from the PdPhoto royalty free public domain stock photos and can be viewed at their website: http://www.pdphoto.org

Save your work when you are finished

Lesson 8 – 5   Using the Picture Tools

When working with pictures in your document you need to decide how you also want the text displayed in relation to your pictures. You can have the text appear under the picture, to the side of the picture, or both. You may also want the text to wrap around the picture.

Open the document called Mount Rushmore1

This file can be found with the downloaded files. This document has a picture of Mount Rushmore along with some text taken (with permission) from the South Dakota Tourism website.

Click on the picture with the mouse

When you click on the picture, a new tab will appear just above the Ribbon on the Title Bar. This is the Picture Tools Tab and is shown in Figure 8-16.

Figure 8-16

Click on the Picture Tools tab

This will allow the Picture Tools to be brought to the surface. This is one of the tabs that are displayed only when needed. It will only be available if the focus is on the picture. The picture has the focus when you click on the picture. The actual Picture Tools tab is shown in Figure 8-17.

Figure 8-17

Under the Arrange group, click on Text Wrapping

The text Wrapping Drop Down list will appear allowing you to select the style of Text wrapping that you would like to use. The different choices are shown in Figure 8-18.

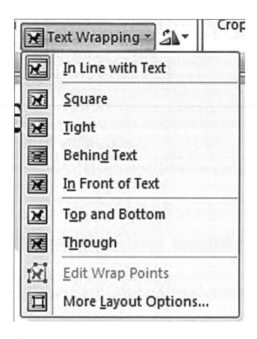

Figure 8-18

Click on the various choice to see how they affect the text wrapping around the picture.

Close this document and open Mount Rushmore2**, also with the downloaded files**

Notice that in the second Mount Rushmore document the picture is not on the far right side of the page. It has been inserted more to the center of the page. This allows the text to be printed on both sides of the picture.

Again go to the Text Wrapping Command and try the various choices to see how they look in the document.

When you have finished, move your attention to the far right side of the Picture tab. We will spend a few moments looking at the rest of the Picture Tools tab.

On the far right side you have the Adjust group. This group, shown in Figure 8-19, allows us to:

Change the brightness and contrast. We can also change the overall color of the picture. Pictures can increase the file size of your documents. To save room on your hard drive, and reduce the time for the document to be downloaded, you can compress the picture. This will allow you to reduce the resolution without losing any of the quality.

Clicking on the Change Picture choice will bring the insert picture dialog box to the screen and allow you to select another picture to insert instead of the current picture.

207

The Reset Picture choice will remove any and all of the formatting that you have done to the picture and restore all of the original formatting (just as it was when you inserted it). If you had clicked on the picture and used one of the sizing handles to resize it, clicking the Reset Picture button would change the size back to the original size.

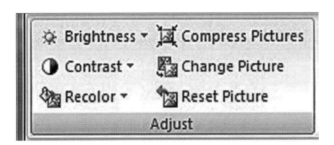

Figure 8-19

The next grouping is for the picture styles and is shown in Figure 8-20.

Figure 8-20

This will allow you to change how the picture is presented on the screen.

Move your mouse over the different styles to see how they affect the picture.

Clicking on the down arrow, shown in Figure 8-20, you can see the next screen of styles.

Also in the Styles Group, on the right side, are drop down lists for picture shapes, picture borders, and picture effects.

Clicking on the Picture shape list will bring the Insert Shapes dialog box to the screen. From here you can actually change the shape of the picture to any of the available shapes.

Click on a few of the shapes to see what happens to the picture.

Don't forget you can use the Undo button to remove any changes that you make.

Click on the Picture Border drop down list

This will allow you to put a border around your picture and decide what color it will be.

Move your mouse over some of the colors and watch the border change colors

The weight choice will let you decide how thick the border is around your picture. The Dashes choice will allow you to use a solid line or use one of the other choices such as the asterisk or a series of dashes.

We have covered part of the Arrange group, but there is more there that just text wrapping.

Click on the Position Button

This will allow you to position the picture in various places in the document in relation to the text.

Move your mouse over the different choice to see how they will change your document

The align button will let you align the picture to the right or the left or center align the picture with the page. You can also align to the top or bottom or to the center of the page.

The Rotate button will allow you to rotate the picture to the right or to the left by 90^0. It will also allow you to flip the picture vertically and/or horizontally.

Click on the Rotate Button and move the mouse over the choice to see the affect they have on the picture

The Size group will allow you to change the size of the picture and crop the picture to the size you want.

The size group is shown in Figure 8-21.

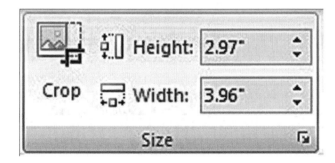

Figure 8-21

If you click on the Crop button the picture will have the crop handles around it as shown in Figure 8-22.

Mount Rushmore National Memorial

features the faces of four George Washington, Roosevelt, and Abraham Hills provide the backdrop world's greatest mountain faces, 500 feet up, look out birch, and aspen in the

This epic sculpture exalted American presidents: Thomas Jefferson, Theodore Lincoln. South Dakota's Black for Mount Rushmore, the carving. These 60-foot high over a setting of pine, spruce, clear western air.

Figure 8-22

You can position your mouse on any of the dark lines and click the left mouse button and drag the mouse to change the size of the picture. Using the crop to make the picture larger will not do the same thing as using the sizing handles as we discussed in Lesson 8-3. If you try to make the picture larger using the crop, will give you a white space between the picture and the newly created larger size. Figure 8-23 shows an example of this.

Mount Rushmore National Memorial

sculpture features the faces American presidents: George Thomas Jefferson, Theodore Abraham Lincoln. South Hills provide the backdrop Rushmore, the world's mountain carving. These 60-500 feet up, look out over spruce, birch, and aspen in western air.

This epic of four exalted Washington, Roosevelt, and Dakota's Black for Mount greatest foot high faces, setting of pine, the clear

Figure 8-23

You can resize the picture by clicking on the small arrows on the Height and/or Width buttons.

Click on the upward pointing arrow on the Height button and watch the picture change to a larger size

You should also notice that the width is getting larger as you make the height larger. The height and the width are tied together so that when one changes the other will also change. This way you will not lose any quality when you change the size. When you use the sizing handles discussed in Lesson 8-3, you can change the height or width individually. Using the height and width button here, you change them both at the same time.

You should also note that there are a few dialog box launchers located on the Picture Tools tab. Clicking the dialog box launcher will give you more options under each section.

Lesson 8 – 6   Working with Charts

Another way to make your documents look professional is to add charts to your documents. In this lesson we will add a chart to our Branson Homeowners Association presentation.

Open the Homeowners Association **document**

Press Ctrl and End to get to the end of the document

Add a new page to the document

Add the headings

Branson Homeowners Association

Projected Profits

Change the style to match the previous headings

Charts are a little more difficult to work with than anything we have done yet. We not only have Word that we are working with, but we will also be working with Microsoft Excel, a spreadsheet application. When we insert a chart into our document, we have to establish each value for the columns and rows (the X and the Y axis) or if we use a pie chart we have to establish the value for each section of the pie. In our project we are going to use a simple bar chart to show the projected income for three years. Are you ready? Here we go!

Make sure the insertion point is at least one line below the headings

On the Insert tab click on chart

A list of the various chart types will be presented in the form of a dialog box. This dialog box is shown in Figure 8-24.

For our chart we will use the Clustered Cylinder. Having a round column will look a little better than the regular old square column.

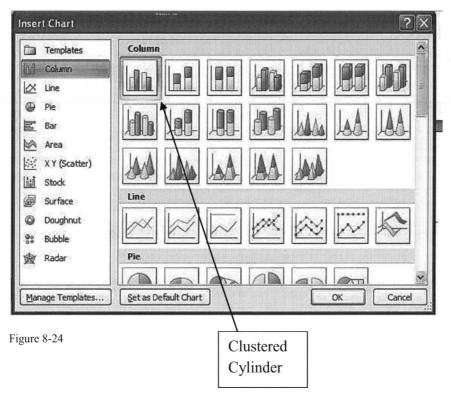

Figure 8-24

Clustered
Cylinder

Click on the Clustered Cylinder and then click OK

This is where the fun starts. When you clicked on the OK button, an Excel spreadsheet jumped onto the screen. We will use the spreadsheet to adjust the number of columns and the values each column represents. The spreadsheet is shown in Figure 8-25. This figure will also show each section we will adjust.

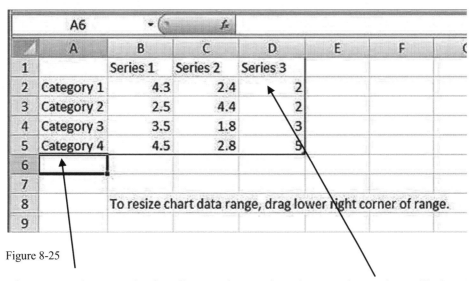

Figure 8-25

The categories are the heading under each column. The series will determine how tall each column goes on the chart. The first thing we will want to do is change the name that goes under the columns to something that makes sense for out document.

213

Click your mouse in cell A2 (the one with Category 1 in it), and type 2007

Repeat this process for cells A3, A4, and A5 using the text 2008, 2009, **and** 2010 **respectively**

You can click on each cell to move to it, or you can press the down arrow on the Keyboard, or you can press the Tab key to move across and then down, but this is harder than just clicking on the cell with your mouse or using the down arrow key. The change you make will not take effect until you click outside of the cell.

When you have finished this part of the changes, your spreadsheet should look like Figure 8-26.

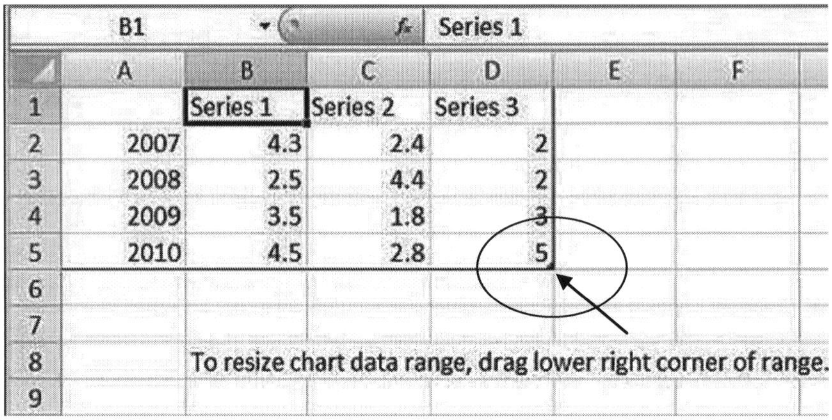

Figure 8-26

The next thing we need to change is the text Series 1. We will want our chart to show that we are looking at our Income.

Change the text in Cell B1 to be Income

We will not be needing columns C and D in our chart. If we were looking at more than one thing we would need other references. If we were comparing income and expenses, we would have two columns, one representing each.

Click the lower right corner of the range of cells (where the arrow is pointing in Figure 8-26) and drag the mouse to the left until columns C and D are highlighted in gray, and then release the left mouse button.

The spreadsheet should now look like Figure 8-27.

	A	B	C	D	E	F
			C	D	E	
1		Income	Series 2	Series 3		
2	2007	4.3	2.4	2		
3	2008	2.5	4.4	2		
4	2009	3.5	1.8	3		
5	2010	4.5	2.8	5		
6						
7						
8		To resize chart data range, drag lower right corner of range.				
9						

Column Headings

Figure 8-27

Even though columns C and D are still in the spreadsheet, they will not appear in our document. Only the cells that are outline in the blue will be in the chart. Just so we don't confuse you because you can see them, let's delete columns C and D so you won't be able to see them.

Select columns C and D with the mouse

You can select the columns by clicking, and holding the mouse button down, on the letter C in the column headings and dragging the mouse to the right until the column D is also highlighted. When both columns are highlighted, release the left mouse button.

When both columns are highlighted, press the Delete key on the keyboard

This should cause all of the text in both columns to be deleted. Your spreadsheet should look like Figure 8-28

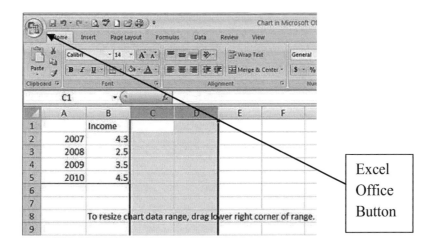

Excel Office Button

Figure 8-28

215

Now we should see what it is going to look like in our document.

Click the Excel Office Button (as shown in Figure 8-28) and then select Close

This will close the spreadsheet and put the chart into the document. The result is shown in Figure 8-29.

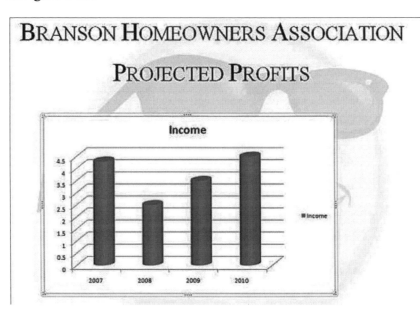

Figure 8-29

This looks pretty good, but I am not sure that I want to show such a large drop between 2007 and 2008. Let's change the data, so the profits look a little better. Hey this is only make-believe, we can manipulate the data any way we want.

Right click inside the body of the chart

This will bring the shortcut menu for the chart to the screen. The menu is shown in Figure 8-30

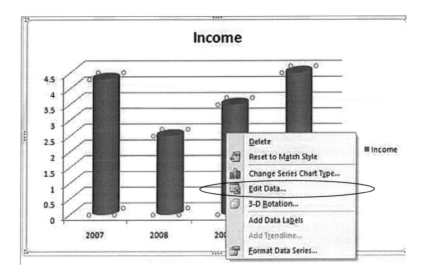

Figure 8-30

Click on Edit Data from the shortcut menu choices

This will bring the spreadsheet back to the screen and allow us to edit any data we want.

Change cell B2 to: 2.5

Change cell B3 to: 2.7

Change cell B4 to: 3.1

Change cell B5 to: 3.7

If you cannot remember how to change the data inside a cell, go back and check how you changed the headings under to columns.

The spreadsheet should look like Figure 8-31

	A	B	C	D	E	F
1		Income				
2	2007	2.5				
3	2008	2.7				
4	2009	3.1				
5	2010	3.7				
6						
7						
8		To resize chart data range, drag lower right corner of range.				
9						

Figure 8-31

217

Close the spreadsheet as you did before

The chart should now look like Figure 8-32.

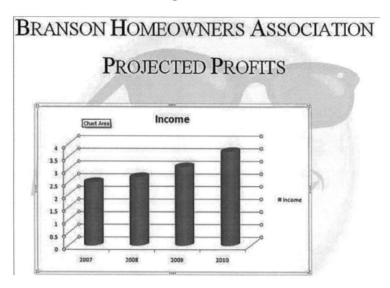

Figure 8-32

This is looking better, only a few minor changes to go. The chart needs to be centered on the page. We also need to explain what denomination the income is based on. 2.5 looks good, but 2.5 what?

Click on the chart and then click the center align button on the Home tab in the paragraph group

This will center the chart on the page.

Click on the Design tab and choose Edit Data in the data group

This will bring the spreadsheet back to the screen.

Change the text in cell B1 from Income to Income in Millions

Now the chart should make more sense.

On the Design tab you can also change the color of the cylinders, by clicking on one of the chart styles.

From the Layout tab, you can insert picture, shapes, and text boxes into the chart area. You can also change how the titles and labels are displayed.

Just for practice, let's insert a shape into out chart.

Click on the Layout tab and then click on shapes in the Insert group

Find the Happy Face under the Basic Shapes group and click on it

When you click on the Happy Face, the mouse cursor will change to a plus sign.

Move the plus sign above the legend "Income in Millions" **and click the mouse button**

If the Happy Face is not positioned as shown in Figure 8-33, click on it with the mouse and drag it to the desired location in the chart.

Save your work

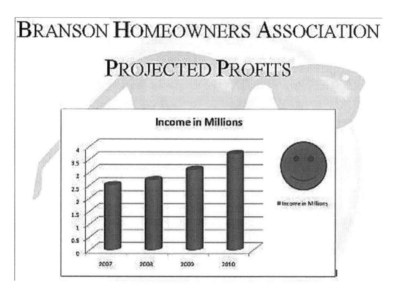

Figure 8-33

Chapter Eight   Review

There are two orientations for the paper, Portrait and landscape. Portrait has the long sides of the paper going up and down while landscape has the long sides of the paper at the top and bottom. The orientation can be changed by clicking on the Orientation command that is in the Page Setup Group which is part of the Page Layout Tab of the Ribbon.

A watermark can be inserted into every page of a document. If you do not want to use one of the standard watermarks you can create your own custom watermark. Your custom watermark can be text or a picture. If you choose to use a picture as your watermark, you may need to use the "washed out" effect to prevent the watermark from being too bright. Some picture will not require this.

A Textbox will allow you to provide additional information to the viewer. A Textbox will NOT automatically adjust to allow the text to fit inside of it.

You can insert shapes, symbols, and clipart into your documents.

Pictures can also be inserted into your documents. The can be rotated to add an extra effect to your document. You use the Free Rotate Handle to rotate the pictures.

Text can be set to wrap around pictures as well as be beneath, in front of, or behind the picture.

Borders can also be added to the pictures.

Charts can be inserted into your documents. This will tie two programs together: Word and Excel.

Chapter Eight   Quiz

1) In which tab and group is the watermark command found?
2) You can set your own watermark text if you desire. **True or False**
3) A Textbox will automatically adjust to accommodate any text typed. **True or False**
4) The command to add a shape to your document, would be found on which Tab?
5) Once inside a document, ClipArt cannot be moved or resized. **True or False**
6) In what group and tab would you find the command to change the brightness of a picture in your document?
7) If you crop a picture and make it smaller, the new picture will have the entire original picture still there and only the size will change. **True or False**
8) What other program works with Word when you insert a chart into a document?

Chapter Nine   Hyperlinks

Now we are going to try something that most people would like to use in their documents, but they have no idea how to make this happen. I am sure you have seen a document (or a web document) that has used a hyperlink. A hyperlink can be a section of text (or even a word) or it can be an object (such as a picture) and when you click on it you are directed to a different section of the document or even to a new document. In this lesson we will create a hyperlink to take us back to the beginning of our Homeowners Association document.

Lesson 9 – 1   Creating a Bookmark

A bookmark identifies a location or a selection of text that you name for future reference. For example, you could have a description of today's lunch menu at the very end of your document. The document itself could be several pages long and the lunch menu is mentioned several times. You could want the person reading the document to be able to click on the words "lunch menu" and the menu to be displayed on the screen. The lunch menu would have a bookmark on it, so Word would know what to bring to the screen when the lunch menu was clicked.

Open the Homeowners Association **document**

Press Ctrl and Home to get to the beginning of the document

Select the text Branson Homeowners Association

This can be done by clicking the mouse in the left margin of the first line or you can click at the beginning of the first line and drag the mouse over the text.

Once the text is selected we need to identify it as a book mark.

On the Insert tab and under the Links group, click on Bookmark

The Bookmark Dialog Box will appear on the screen. All we have to do now is give the bookmark a unique name. This name will keep this bookmark separate from any other bookmarks we might create.

The Insertion point should already be in the name textbox. If the insertion point is not in the name textbox, click on the textbox.

Type the word Top **in the textbox**

The dialog box should look like Figure 9-1

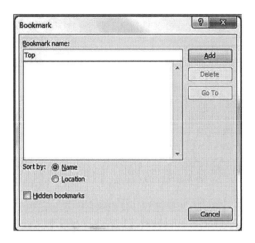

Figure 9-1

Click the Add button

This will create the bookmark that we need.

Lesson 9 – 2   Creating a Hyperlink

Now that we have the bookmark set in place, we need to create a link between the two sections. This link is called a hyperlink. The text we will use for this link is called Hypertext, which means "more than just text".

Press Ctrl and End to navigate to the end of the document

The end of the document is probably going to be immediately on the right side of the chart. We will want to put two blank lines between the chart and the hyperlink.

Press Enter until the mouse is two lines below the chart

Center align the text

Change the text size to 9

If we left the text the same size as the document text, it wouldn't look quite right. The hypertext is usually smaller than the regular text in the document.

Type the following text   Back To Top

Select the text you just typed

You must select the needed text before you can assign it as a hyperlink.

On the Insert tab, click on Hyperlink

The Hyperlink Dialog Box is brought to the screen. This is where we will tell Word where the other end of the link is located. The Hyperlink Dialog Box is shown in Figure 9-2.

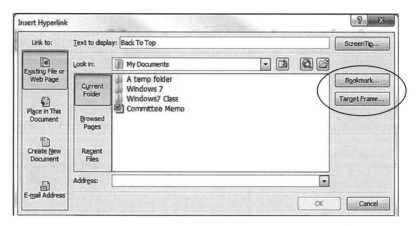

Figure 9-2

225

If we use the "Look in" list box we will be connecting the link to another document. Since we want to stay in this document, we will choose the Bookmark button.

Click on the Bookmark button

This will bring another dialog box to the screen. This dialog box will allow us to set the link to the bookmark we want to use. There could be several bookmarks in our document and we would need to tell Word which one to use. Our choice is going to be easy; we have only created one bookmark. Figure 9-3 shows the new dialog box.

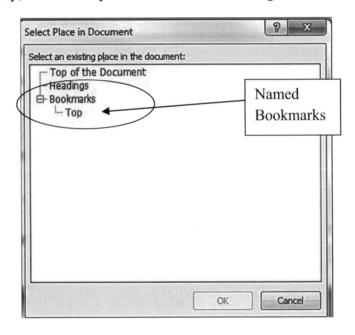

Figure 9-3

Click on the Bookmark named Top and then click OK

This dialog box will disappear and the first dialog box will remain on the screen. To set the bookmark we must also click OK in this dialog box.

Click OK

The hyperlink is now created. Notice that the text "Back To Top" is now underlined in blue. This indicates that the text is now a hyperlink. Now exactly how do we use this hyperlink?

Lesson 9 – 3   Using the Hyperlink

I stated earlier that the user has to click on the hyperlink to move to the linked text. Well actually the user must do one other thing; they must hold the Ctrl key down and then click the mouse on the link.

Move the mouse over the hyperlink and watch the screen tip

The screen tip, as shown in Figure 9-4, will show you the name of the bookmark it is linked with and instructions on how to use the hyperlink.

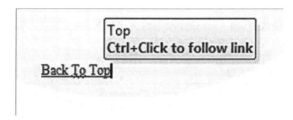

Figurer 9-4

Hold the Ctrl key down and click the mouse on the hyperlink (the text Back to Top) then release the Ctrl key

You will be taken directly back to the first page of the document to the text Branson Homeowners Association. Now you can create a link from any place in the document to anyplace else in the document.

Save any changes made and close the document

Chapter Nine   Review

Using a hyperlink will allow the user to click on a section of text (or an object) and be redirected to another location in the document or to a different document.

A bookmark is the final destination of the hyperlink, if you stay inside the current document. All bookmarks must have a unique name.

To create a hyperlink, select the text or object you want the user to click on and then click on the Hyperlink command that is on the Insert Tab. You can determine where the hyperlink destination will be from the dialog box that comes to the screen.

To use a hyperlink, the user must hold the Ctrl key down and then click the mouse on the link.

Chapter Nine   Quiz

1) A _____ identifies a location or a selection of text that you name for future reference.

2) Before text can be assigned as a hyperlink, you must select the text. **True or False**

3) All bookmarks must have a unique name. **True or False**

4) A hyperlink can direct the user to an existing file or web page. **True or False**

5) When you click on the Bookmark button that is in the Insert Hyperlink Dialog Box, Bottom of the Document is already a defined bookmark. **True or False**

6) If you move the mouse over a hyperlink and let it hover there for a few seconds, a _____ _____ will appear with instructions on how to use the hyperlink.

7) Explain how to use a hyperlink.

Chapter Ten   Using Mail Merge

Mail merge is a wonderful feature of Word. What if you wanted to send a letter to several people but you wanted each letter personalized with Dear whoever at the top? You could write everyone the same letter and put each person's name at the top. That wouldn't be very much fun and would take a long time. You could write one letter and print it and then go back and change the name at the top and print it again. Well if you only had one or two letters that might not be so bad.

What if you had to send out an e-mail (or a letter) to 150 employees showing their income and tax withholdings? Every letter would have to have the employees name and address. It would also have to have each employee's social security number and income and the amount of taxes withheld. Now this could take hours.

Why not let Word do all of the hard stuff for us? We could type the letter that has the information that is the same for everyone and put placeholders in the document for all of the information that is unique for each individual. Word will take care of the rest.

This entire chapter will be devoted to making this happen. Mail Merge is a long process and it is somewhat complicated, but together we can do this. So, go get a cold drink of water and let's get started.

Lesson 10 – 1   The Basics

For Mail Merge to work there are two things you must have. The first is a master document. This is the letter that everyone will receive. The second thing is a data source. This is a place where all of the unique data is kept for each person receiving the letter. When you do the Mail Merge, you will have a unique document for each person. Figure 10-1 might explain this concept a little better.

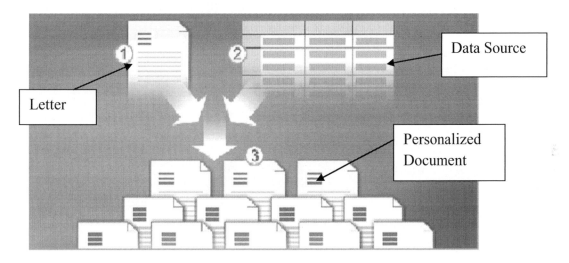

Figure 10-1

I want you to understand this part so you will know what to expect when we finish.

1) This is the main document that everyone's letter will start with. Almost everything in this document will be the same for everyone.
2) This is where we will store individual information such as each person's name and address. This part of the document will be different for each person.
3) Word takes the letter and the data source and combines the two elements to make a unique letter for each person. At the top of each letter would be each person's name and address (or whatever is in the data source) and would be different for each letter. The body of the letter will be the same no matter whose name is at the top.

In the rest of this chapter we will create a silly Christmas letter and personalize it for a few family members.

The main document is where we need to start. This is the letter we plan on sending everyone. This document will contain the body of the letter and placeholders for the address and greeting name. This part will only have to be done once no matter how many letters you plan on sending out.

Microsoft must have realized that people do not do mail merges every day, because the created a very helpful tool in the mail merge section of the Mailings tab of the Ribbon.

Create a new document

This can be done by clicking on the Office button and selecting new and then blank document.

Click Start Mail Merge on the Mailing tab of the Ribbon

When you click this button, a drop down list will appear on the screen. This list is shown in Figure 10-2.

Figure 10-2

On the drop down list select the Step by Step Mail Merge Wizard

This is the wonderful tool I was telling you about. Word will guide us through the process of creating a mail merge. The first thing that will happen is the Mail Merge Pane will jump on the screen. This is what will guide us through the process. The pane is shown in Figure 8-3.

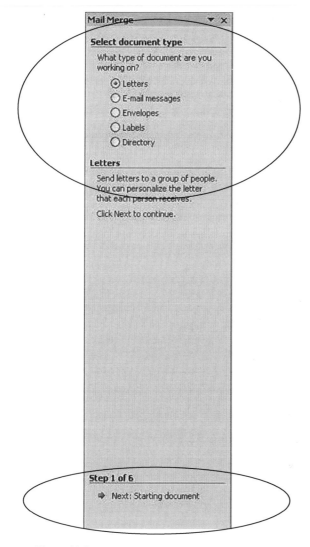

Figure 10-3

The first thing Word wants to know is what type of document we are working on. For our choice, we are going to be typing a letter. Since this is already checked we don't have to do anything. If we were creating an e-mail, we would click in the radio button next to e-mail.

Down at the bottom it is telling us that this is step one of six steps.

Click Next down at the bottom

When you click the next link, a new set of choices will replace the original choices. This is shown in Figure 10-4.

Figure 10-4

The choices are: Use the current document as our main document, start our document from a template, or use an existing document as our main document. Since we just created a new document for this purpose, let's use the current document. Again, this choice is already selected, so we don't have to do anything.

Down at the bottom we can choose to move on to the next step or return to the previous step.

Lesson 10 – 4   Step 3

Click on Next

When you do, new choices will be on the pane. These are shown in Figure 10-5.

Figure 10-5

With this choice, we need to decide who our recipients for the letter will be. We can use an existing list of recipients, or use our contact list from Outlook. If neither one of these choices will work, we can create a new list of recipients.

Click on the third choice: type a new list and click create, which will appear in the middle section

We chose this choice because we don't have a list created yet, also you are doing this to learn. We need to make a new list, so you can learn how it is done.

This choice will cause a new dialog box to come to the screen, as shown in Figure 10-6.

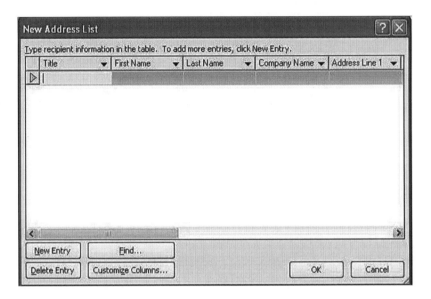

Figure 10-6

If you use the scroll bar at the bottom to view the choices across the top, you will see that there are several entries that we do not need. For our Christmas letter we will not need a company name and we probably won't need two address lines. We probably won't need the phone number printed or the country or the e-mail address on the Christmas letter. Since we won't need these, let's customize the columns.

Click on Customize Columns

This choice will allow us to add columns or delete them as needed. We could also change the name (heading) of the columns.

Clicking Customize will bring another dialog box to the screen. I will show you this dialog box in Figure 10-7.

Figure 10-7

Click on Company Name under field names and then click Delete

When you click on delete Word will ask you if you are sure you want to delete this field. If you choose to delete the field and there was any information contained in this field, it will be deleted along with the field.

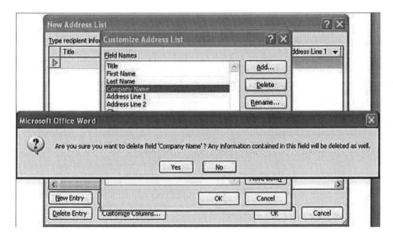

Figure 10-8

This is a new list with no information so we can delete the field without losing any data.

Click on the Yes button

Delete the following fields:

> Address Line 2

> Country

> Home and Work phone

> e-mail address

Click OK to go back to the first dialog box

> Now we need to add data into the fields.

Under Title type Mr.

Press the Tab key to move to the next entry point

Under First Name type the first name of someone you know

Under the Last Name type the last name of the person you decided to use in the first name

Press the Tab key and fill in the rest of the information for the first person

> If you press the Tab key when you fill in the last entry, you will start a new entry.

Add at least two more entries to your list

> When you have finished adding entries to your list, you will need to save the list.

Click OK to start the save process

> Word needs to know where you want to save the file. Word has to know this to be able to find the file later to finish the mail merge.

> When you click the OK button, another dialog box will come to the screen, the Save Address List Dialog Box. This is shown in Figure 10-9.

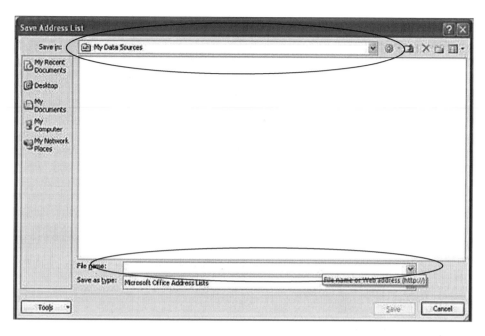

Figure 10-9

We could keep the file in the default place, My Data Sources, or we can change where the file is located. Just so you can get use to using the Save Address List Dialog Box, let's save it in a different location. Let's save it in the Word 2007 folder.

Navigate to the Word 2007 folder

Use the drop down arrow in the Save In section to find the Word 2007 folder you created way back in the beginning (probably under your documents). After you find the folder, you have to give the list a unique name. For our lesson we can just call it Test List.

Click in the File Name text box and type Test List **and then click Save**

Guess what? Another box comes to the screen. Will this madness never end? We are almost finished, so hang on. The new box is shown in Figure 10-10.

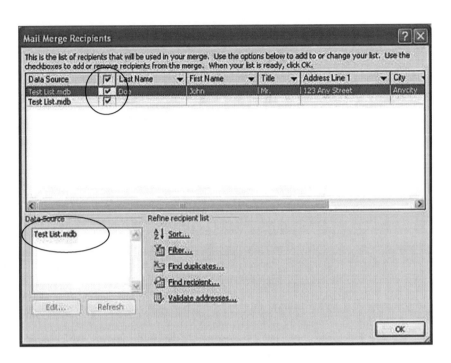

Figure 10-10

From this dialog box you can choose which recipients in the list will receive the letter. If you had one large list, such as your complete e-mail list, you might not want everyone on the list to receive this Christmas letter. You can click the checkbox next to a name to include them in the list if it is not already checked. If there is a check mark next to the name and you click it, that name will be removed from getting the letter. If you click the checkbox at the very top, everyone in the list will be included as a recipient.

If you need to edit the data source, you can click on the name of the data source in the data source box at the bottom. After you click the name of the data source the edit button will become available for you to click on, so you will be able to add or delete names from the list.

Everything should be fine with this list, so click OK

Now we can finally write our letter.

Lesson 10 – 5   Step 4

On the pane, down at the bottom, click on Next: Write your letter

The Mail Merge Pane contents will now change providing help on writing your letter, as shown in Figure 10-11.

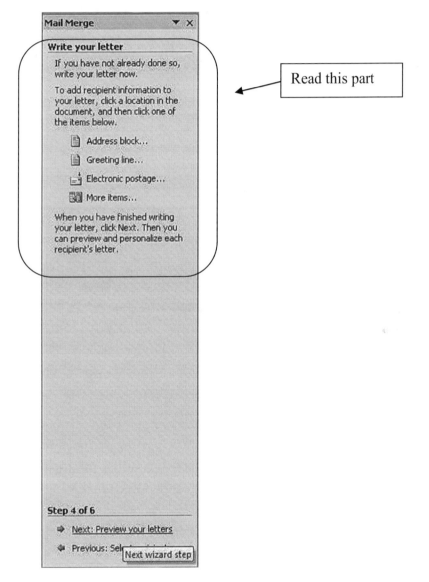

Figure 10-11

Next we will insert some of the most common things that will go into a letter. These things may not be what you would normally put into a Christmas letter, but it is good practice.

In your blank letter, click on the top line

Next click on Address Block in the Mail Merge Pane

Bet you can't guess what will come on your screen. Okay, it is another dialog box. Wow! Was that a surprise or what? See Figure 10-12 for the dialog box.

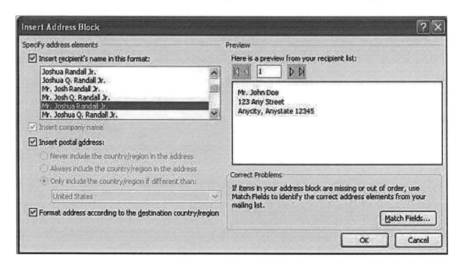

Figure 10-12

With the Insert Address Dialog Box you can change the way the information is displayed in the address block. You will notice that the format is shown on the left and the actual data from the first person in your list is displayed on the right. You can click on the different format styles and watch the results on the right to see which format you want in the letter. If you uncheck the Insert recipients name checkbox, the name will be removed from the address block. If you uncheck the Insert postal address checkbox, the address will be removed from the address block.

You may want to read the text under the Correct Problems section. If your information, for whatever reason, does not match the format on the left, you can change the order of where in the data source the information is located. This will make a little more sense if you click on the Match Fields button.

Click on the Match Fields button

This will bring up; you guessed it, another dialog box. See Figure 10-13 for the dialog box.

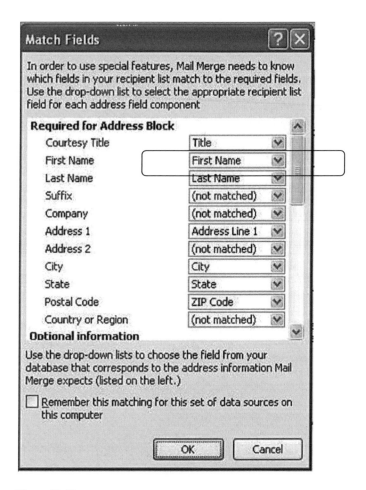

Figure 10-13

Let's say in our hurry to set the mail merge up, we accidentally mixed the last name and the first name up. If we did that the address block would have Mr. Doe John instead of Mr. John Doe. If all of the information was entered this way, we could change where Word will look for the needed information.

Assuming that the first and last names were put in backwards, this is how we would fix everything.

To make the first name use the information that is stored in the last name position of the data source, you would click the down arrow next to the Last Name on the right side. When you click on this, a drop down list will come to the screen, as shown in Figure 10-14.

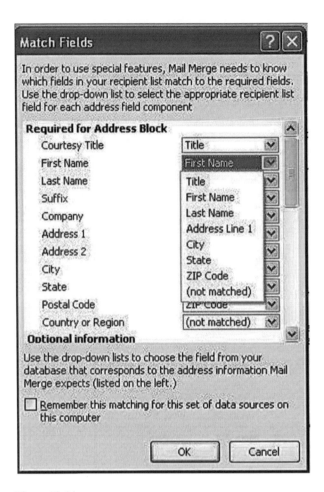

Figure 10-14

The drop down list will allow us to choose where the actual first name data is stored. As you remember, we are pretending that we entered the last name in the first name place and the first name in the last name place. So, all we would have to do is click on Last name and Word would use whatever was in here for our First name. We would repeat this process for changing the last name.

Our data is correct, so click on Cancel

This will take us back to the Insert Address Block dialog box.

Click OK and look at your letter

None of the information appears to be in the letter. The information is held in the data source, remember? What is in the letter is a placeholder for the data that is stored in the data source. You need to remember that this part of every letter will be different, so we put a placeholder in the letter and the actual name and address will be added later. You can tell that it is a placeholder because it has a chevron (<< >>)before and after it.

244

Now let's get on with our letter.

Press Enter twice so we will have a blank line between the address block and the body of our letter

Now we work on the next line of our letter.

We want each letter to have the first name of the recipient after the word dear. To accomplish this we will need to insert another placeholder. This placeholder will hold only the first name of the recipient.

In the Mail Merge Pane, click on the Greeting Line

It's time for more choices. The new dialog box is shown in Figure 10-15.

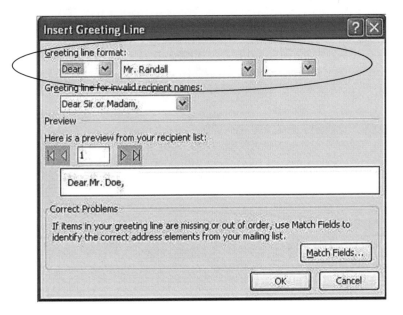

Figure 10-15

From this dialog box we can choose the format of the greeting line. In the first drop down list we can choose Dear, To, or nothing at all. In the second drop down list we can choose how the name is displayed. In the third drop down list we can choose if we need a comma, or a colon, or nothing at all.

In the bottom preview area we are shown what each choice will look like. We do not want this to be as formal as a business letter. It would be better for a Christmas letter to have Dear and the first name of the recipient.

Click the down arrow in the center box (the one that has Mr. Randall in it) and scroll down until you can see the first name only.

Click on the first name (Josh) and then click OK

Now we have a greeting line placeholder in our document. The greeting line of each letter will have a different name in it.

When you are finished, press the Enter key

Now we can type the letter that will be same for everyone.

Starting with the line directly under the greeting line, type the following:

Well it is that time of the year again, time for another letter from me. This is the only time of the year that you will hear from me. Well here is the latest scoop from my house. The kids are all grown and do not have time for me anymore. The economy is bad and I don't have the money to go anywhere. There is nothing new in my love life. My hair is falling out and my stomach is getting bigger. My knees are going bad and will probably need to be replaced.

Hope you have a great holiday season

After you are finished, we will have to go back to the Mail Merge Pane and see what we need to do next.

Lesson 10 – 6   Step 5

Click on Next: Preview your letters (down at the bottom of the pane)

In this part of the mail merge we get to see the actual letters that are going to be sent out. The Mail Merge Pane contents will change and we will have some different options. We can shuffle through the entire list of recipients and view the letter that will be sent to each one of them. A view of the screen is shown in Figure 10-16.

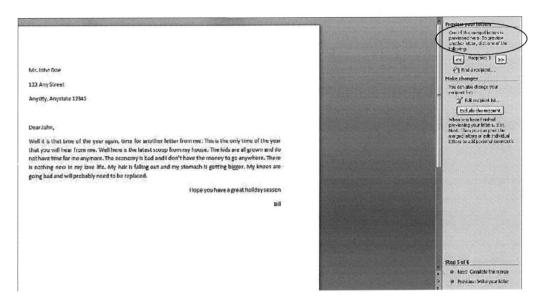

Figure 10-16

In the top part of the mail merge pane, are arrows we can use to scroll through the list of recipients. Each letter will have a different name and address in the address block. Each letter will also have a different greeting line. Now we need to finish the mail merge.

Click the next link to complete the merge

We can now print the letters by clicking on the print letters link on the pane. If we wanted to, we could edit individual letters. This would be great if there were one or two letters we wanted to add a personal note to the letter. If you click on the Edit individual letters link, you can choose to edit all letters, the current letter on the screen, or all letters in a range (but you will need to know the starting and ending point).

That is all there is to a mail-merge. I realize that this seemed to take a long time and it was complicated, but if you had to write a few hundred letters and each one had to be personalized, it could take hours. All you have to do is let Word work for you.

Save your letter in the Word 2007 folder and save it as Christmas Merge

Chapter Ten   Review

Main Merge lets you write one letter and personalize it for several different people. Mail Merge requires two things: a master document and a data source. The master document is the letter (or document) that everyone will receive, and the data source is the place where all of the unique data is kept for each person receiving the document.

The easiest way to create a mail merge is by using the Mail Merge Wizard. This will guide you through the entire mail merge process. Listed below are the six steps of the mail merge.

1) Decide what type of document you are going to use.
2) Decide if you are going to use the current document or an existing document.
3) Decide if you are going to use an existing list of recipients or create a new list.
4) Write the letter.
5) Preview your letter.
6) Complete the merge.

Chapter Ten   Quiz

1) The Main Document, the letter you are sending to everyone, contains the body of the letter and _____ for the address and greeting line.

2) The _____ _____ _____ will provide you with step-by-step instructions for creating a Mail Merge.

3) A letter is the only type of document that works with Mail Merge. **True or False**

4) You must create a new list of recipients for every mail merge. **True or False**

5) When you are in the New Address List Dialog Box and you want to add or delete columns, you click the mouse on which button in the dialog box?

6) If you are deleting a column in the address list, the column will be removed from the list but the actual data will be preserved. **True or False**

7) When saving an address list, it must be saved in "My Data Sources". **True or False**

8) Using the Insert Address Block Dialog Box will allow you to change the way the information is displayed in the Address Block. **True or False**

9) What is the name of the dialog box that will allow you to change the format of the Greeting Line?

10) In the last step of the mail merge, you can edit individual letters if you desire. **True or False**

Chapter Eleven Before you Print

This may sound like a useless chapter. After all what is the big deal? You either print or you don't, right? We have covered the basics of printing back in Chapter one. In this chapter we will go beyond simple printing. We will deal with some of the other things that you will want to consider having in your document before you decide to print it.

Lesson 11 -1   Cover Pages

Before you print your document, you might want to consider having a cover page for the document. You have to insert the cover page, so it only makes sense that it is on the Insert Tab of the Ribbon. It is located in the Pages Group and is shown in Figure 11-1.

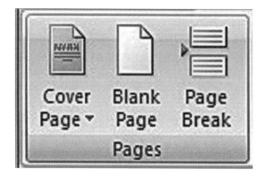

Figure 11-1

Open the Testimonial **document**

This is a document that we have used before and we can insert a cover page into the document.

Click Cover Page on the Pages Group

A list of the pre-made cover pages will drop down with a picture of what the pages look like. To insert one into the document all you have to do is click on it. Figure 11-2 shows the drop down list.

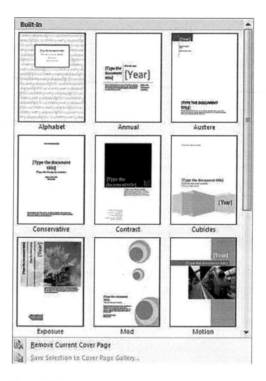

Figure 11-2

When you insert a cover sheet you will have to add certain things. These include the Title you want on the page, the author's name, possibly a date, and perhaps a short summary telling what the document is about.

Click on the Puzzle cover page

You will notice that for the Title Word put the first line of our document. This is not a very good tile for our document, since it does not tell us anything about the document. The title should be something that identifies the document or at a minimum makes sense.

Let's change the title to Testimony. As you move the mouse over the title, the title area changes color. When it changes color we can click on it and then change the title.

Click and drag the mouse over the entire title

We want to select the complete title so that when we start typing a new title the old title will be replaced. If we don't select the title then we have to manually delete the other title.

Type the following title

Testimony

When you are finished typing, pressing the enter key will not move you to the next level so you have to manually select the next choice.

Click on Subtitle and press the space bar

We don't have a subtitle for our document but if we press the delete key it will not get rid of the text. We will use the spacebar to replace the existing text with a space.

Click on the abstract and type the following text

This is my testimony.

Click on the Year and click on the down arrow and choose today

Down at the bottom make sure your name is on the right, do not put a company name (unless you want your company name on the document), and make sure today's date is below that

This is a great way to add a cover sheet to your document, but what if you don't see one that you like? We could create our own cover sheet and use it.

Close the document and do not save the changes

Reopen the Testimonial **document**

Now that we are back to the original document we can design our own cover sheet.

Make sure that the insertion point is at the beginning of the document

Insert a blank page

This is done by clicking on the Insert tab and then clicking on the Blank Page icon in the Pages Group. Figure 11-3 shows the Pages Group.

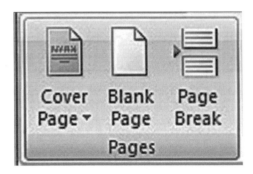

Figure 11-3

A blank page will be inserted into the document just before the insertion point. This will give us a new page that we can use to design our cover sheet.

Before we start typing our cover page there are a few formatting changes we need to make.

Press Ctrl and Home to get to the beginning of the document

Make the following formatting changes

Font = Times New Roman

Font size = 24

Center aligned paragraph

Font color = Red

Press Enter three times to move the text down a little

Type My Testimonial

Move the insertion point down to the bottom right corner and type your name (make sure the text is right aligned and the font is no larger than 12)

Granted this is not a very good cover page, but for what we are doing it will work. We may want to use this design for another cover page so we are going to save it.

Before you can save the cover page you must first select the text you want to save. If you do not select the text, the save choice will not be available.

Select the entire first page of the document (the one we just made)

Click on Cover Page on the Insert tab

At the bottom click on save selection to cover page gallery

The Create New Building Block Dialog box will come to the screen. All we have to do is give our cover page a name and then we can save it (See Figure 11-4).

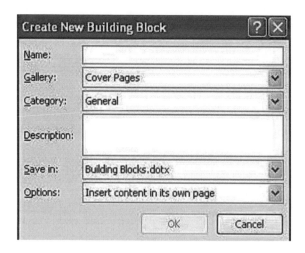

Figure 11-4

Type in the name Test Cover Page **and then click OK**

The cover page will now be saved with the rest of the cover pages and we can use it whenever we want.

Open the cover pages and verify that our new one is saved

Close the document

You can save the changes if you want.

Lesson 11 – 2   Page Breaks

When you are typing a document the page breaks do not always fall where you would like them. In this lesson we will learn how to insert a page break into your document.

Open the BranTel Info3 **document located in the files that you downloaded**

As you can see, the end of the first page is right in the middle of a paragraph. The document would look much better if the entire paragraph was together on one page. We can fix that.

Move the insertion point just before the word Support

In the Pages Group of the Insert Tab is a command that inserts a page break. It is named appropriately Page Break.

Click the Page Break Command

The page break is inserted into the document just before the insertion point. Now the entire paragraph is together and it looks so much better.

If you change your mind, you can use the Undo button to undo the page break. With the undo you can choose which action you want to undo.

Click on the down arrow of the Undo button and choose page break for the drop down list
(See Figure 11-5)

Figure 11-5

Close the document

Lesson 11 – 3   Headers & Footers

Most people don't bother with headers and footers in their documents. This can be a terrible waste of resources. Just think you can have your company name at the top of every page of every document that you send out. If you are in business, you want your company out there as much as possible. Perhaps you are like me and you just want to see your name in print all of the time. Either way, headers and footers are something you should consider.

Open the Computer Lab **document that is with the downloaded files**

We will insert a header into this document.

Since this is something we will insert it is logical to have it on the Insert Tab in the Header and Footer Group (See Figure 11-6).

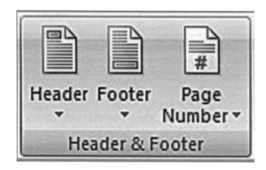

Figure 11-6

Click on the Header command

A list of the available headers will drop down.

Scroll down until you find the Tiles **header and then click on it** (See Figure 11-7)

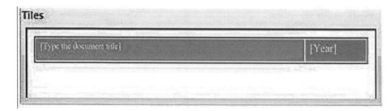

Figure 11-7

The header will immediately be inserted into the document and will be at the top of every page in the document. The text for the header will by default be the top line of the document, in this case Computer Lab Committee. If this text does not suit your needs, you can click on it and change it, perhaps to your company name. As it turns out the name will need to be changed.

Click on the name, at the end, and change the name to match the text below

Computer Lab Committee of BranTel Communications

The year also needs to be changed.

Click on the year and then on the down arrow

Click on today

Click on close header and footer

The header part of the document is now complete, unless you decide that it needs to be changed. If this is the case you can edit the header or remove it completely from the document. To edit the header you need to click on the Header command as you did before and chose Edit from the bottom of the header drop down list (See Figure 11-8), you could also choose remove if you wanted to remove the header.

Figure 11-8

You might want to consider adding a footer to your document as well. A footer is like a header only it is at the bottom of the page. In our document we will want to keep the style consistent so we will choose the same type as the header.

Click on the Footer command

The Footer drop down list will pop onto the screen. As I said before we want the document to have some consistency so we will use the same kind of Footer.

Scroll down to and click on the Tiles type of Footer (See Figure 11-9)

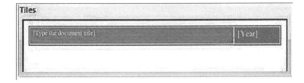

Figure 11-9

Go to the bottom of the page and click on the part that says Type Company Address

Type the following

P.O. Box 1249 Branson, MO. 65615

Click on close header and footer

Our document now has the company department name at the top of each page and the address at the bottom of each page. And let us not forget; it looks cool too.

Close the document and save the changes

Lesson 11 – 4   Table of Contents

Normally one would think that having a Table of Contents in your document would require hours of typing and going back and forth making sure that the page numbers match the titles. If the Table of Contents spilled over onto another page, every page number would have to be changed so that everything would match. If this is what you have done in the past, you will love this.

Open the Homeowners Association **document**

Let's insert a Table of Contents into this document.

One might think that you just need to go to the Insert Tab and insert the Table of contents. Well that's not quite right. The Table of Contents is not on the Insert tab.

The Table of Contents is located on the Reference tab in the Table of Contents Group (See Figure 11-10).

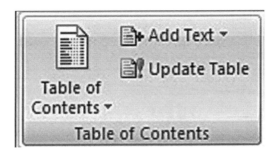

Figure 11-10

Before we insert the Table of Contents we need to decide where we want it to appear in our document. Let's put our Table of Contents after the first page.

Insert a blank page between pages one and two

This can be done by clicking the mouse before the first word on page two and then clicking the blank page command on the Insert Tab in the Page Group.

Click at the top of the new page

Go to the Reference tab and click on Table of Contents

Now we have to decide which style of Table we want to use. I prefer the one that has Table of Contents at the top, so choose the second one down.

Click on the second choice

A new dialog box will come to the screen letting you know that there is nothing to put into the table of contents. There will also be directions as to how you can add items to the table. This is shown in Figure 11-11.

Microsoft Office Word

To add or remove items in the table of contents, select text in your document, and then do one of the following:

* Apply a heading style from the Styles gallery on the Home tab.
* Change the outline level of the text, using the Add Text menu in the Table of Contents group of the References tab.

OK

Figure 11-11

Now the table is ready for us to add items. First we have to select the text that we want in the table of contents and then add the text.

On page three select the text Located in Branson Missouri

Click on Add Text (See Figure 11-12)

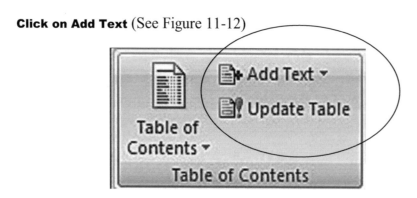

Add Text ▾
Update Table
Table of Contents ▾
Table of Contents

Figure 11-12

Now we must tell Word how we want the text inserted (that is at what level). If we choose level one, the words will be even with Table of Contents at the top. If we had several different things we needed put in the table, such as chapter 1, lesson 1, lesson 1A, this would be advisable. As it is we only have one level, so we can choose level 2 just to offset it from the words Table of Contents

Click on Level 2

Go to the next page and select the text Fun in Branson **and add it at level 2**

We are almost done; we just have to tell Word to include the text we just added. We do this by updating the table.

Click on Update Table

We only have one last choice to make. Do we want to update the page numbers only or do we want to update the entire table? Since we have added information we need to update the entire table.

Click the radio button next to Update entire table and then click OK

Go back to the Table of Contents and check it out!

Add the text Our Newest Venture **on page 5**

Add the text Projected Profits **on page 6**

Close the document and save the changes

Lesson 11 – 5   Table of Figures

In the last lesson we learned how to create a Table of Contents for your documents. In this lesson we will learn how to create a Table of Figures. A table of figures is similar to the table of contents only it lists all of the figures in the document.

Open the Homeowners Association2 **document that is with the downloaded files**

This document is similar to the document from the last lesson. The pictures are slightly smaller to allow room for the captions. The first thing we will have to do is select a picture (or chart or table) so we can add a caption. A caption is necessary to be able to add something to the table of figures.

Click on the picture of the U.S. map

When you click on the picture, your screen should look like Figure 11-13.

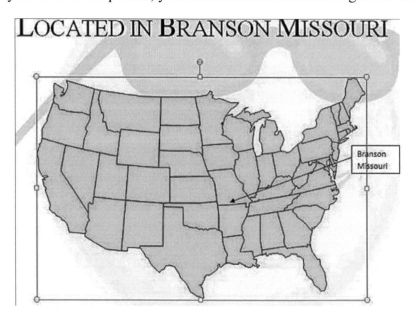

Figure 11-13

Click on the Insert Caption command

This command is located in the Captions Group of the Reference Tab and is shown in Figure 11-14.

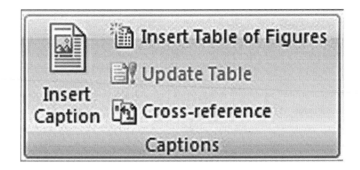

Figure 11-14

Clicking this command will bring the Caption Dialog Box to the screen. The dialog box is shown in Figure 11-15.

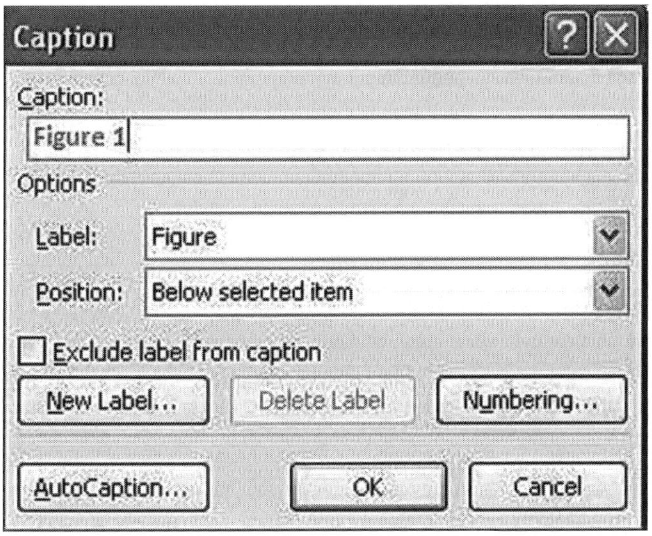

Figure 11-15

There are several things you will want to notice about the dialog box. First, the caption has Figure 1 already in it (it could have also been a Table or Equation). Word automatically identified it as a picture and not a table or equation that is why Word put the Figure 1 in the caption. Obviously this is the first figure and that is why it has the number one in it.

If Word was mistaken in its assumptions you can manually change it from a figure by clicking the drop down arrow across from Label and choosing one of the other options.

The default place for the caption is below the selected item. You can change the position to above the selected item by clicking the drop down arrow across from Position and choosing the above selected item choice.

The insertion point is already in the Caption text box, now we need to fill in the rest of the caption.

Fill in the caption until it looks like the following

Figure 1 – Branson Missouri Location

Click the OK button

The caption will be inserted under the picture and can be used to identify the figure when we add it to the Table of Figures.

Select the second picture and add this caption

Figure 2 – Fun in Branson

Select the third picture and add this caption

Figure 3 – Our Newest Venture

Select the chart and add this caption

Figure 4 – Projected profits

Press Ctrl and End to move to the very end of the document

This is where we will add the Table of Figures.

Click on the Insert Table of Figures command

Guess what? Another dialog box will come onto the screen. This one is the Table of Figures Dialog Box and is shown in Figure 11-16.

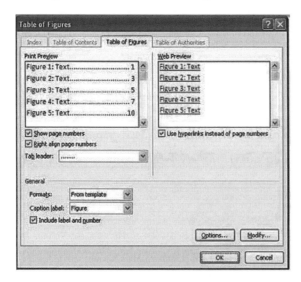

Figure 11-16

There are only a few minor things we need to notice about this dialog box. First there is a preview (well sort of a preview) of what it will look like. Also we are going to show the page numbers and right align the page numbers. We will do both of these so that everything will be neat and organized. We also want to include the label and number, or the Figure 1 part will not show.

On the right side is the Web Preview. Normally most people would not pay any attention to this part because we are not putting our document on a web site. There is a part of this that does affect our document. There is a check mark next to use hyperlinks instead of page numbers. If you leave this checked, a hyperlink will also be inserted into our document. This is a good thing and we want to leave this checked.

Click the OK button

The Table of Figures will be inserted into our document. The hyperlink will also be inserted along with the table. To use the hyperlink you move the mouse pointer over the text and press and hold the Ctrl key and then click the mouse (also let go of the Ctrl key after you click the mouse). You will be taken to the caption of the figure.

Note: You can add new captions and use the Update Table to add additional figures.

Save your changes and close the document

Lesson 11 – 6   Table of Authorities

Some of us, me included, would like everyone to think that we know everything. We tend to state facts in our documents and would be happy if everyone thought that we figured it all out by ourselves. The truth is most of us cannot figure this much stuff out by ourselves. If you were an attorney, you would need to site court cases to prove your point. You may on the other hand just want everyone to know that Mount Elbert has the highest elevation of all of the mountains in Colorado. If you fall into this group of people, you might want to include a table of authorities in your document. A table of authorities lists all of the cases, statutes, and other authorities cited in the document.

Open the Fryingpan **document located with the downloaded files**

This document tells why the Fryingpan-Arkansas project was developed. It is only a few sentences long but it will work for this lesson. In this document we have cited an authority and we will use that in our table.

The first thing you have to do is select the text that you want to use in the table. This should sound familiar, since we did this in both the table of contents and the table of figures.

Select the text Public Law 87-590 (77 Stat. 393)

Now we have to mark the text. We mark the text by clicking on Mark Citation in the Table of Authorities Group which is on the References Tab of the Ribbon (See Figure 11-17).

Figure 11-17

Click on Mark Citation

Clicking on this will bring the Mark Citation Dialog box to the screen as shown in Figure 11-18.

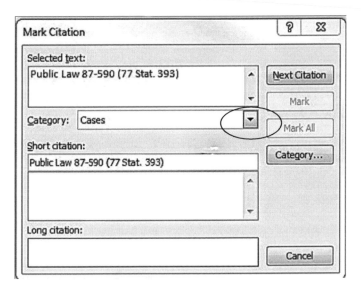

Figure 11-18

In the dialog box the selected text is listed and it is also shown under short citation. These are fine but the category might need to be changed. This is not a court case it is an actual law, so we will need to change the category to Statutes. There are other categories such as rules, treaties, regulations, or other authorities if you have something that does not fall into the preset authorities.

Click on the down arrow in the category section and choose Statutes

Now we need to mark the selected text and we are ready to move on.

Click the mouse on Mark

Right about now your whole screen went crazy with all kinds of stuff that you may have never seen before. Do not panic, we will fix this in about two seconds.

Click the close button

Your screen will probably look like Figure 11-19. Before we fix it (it is not actually broken as you will see) you should know that these things are always in your document but they are hidden from view. Most of the things you are seeing are paragraph markers and dots between each word that represent a space, and they are easily hidden again.

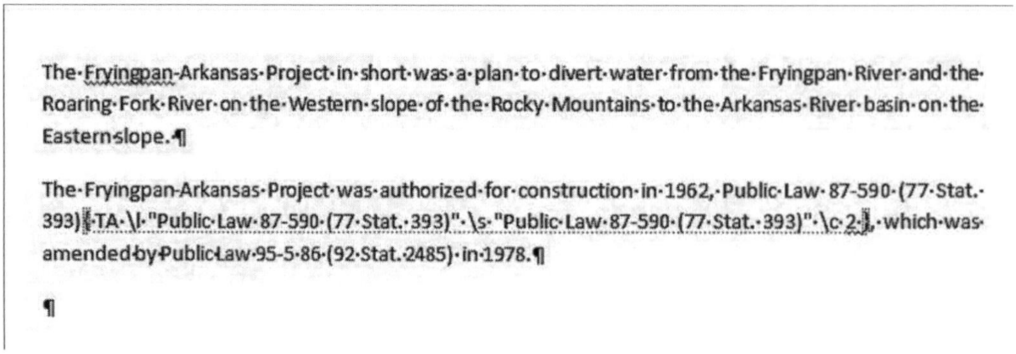

The·Fryingpan·Arkansas·Project·in·short·was·a·plan·to·divert·water·from·the·Fryingpan·River·and·the·Roaring·Fork·River·on·the·Western·slope·of·the·Rocky·Mountains·to·the·Arkansas·River·basin·on·the·Eastern·slope.¶

The·Fryingpan-Arkansas·Project·was·authorized·for·construction·in·1962,·Public·Law·87-590·(77·Stat.·393)⟦TA·\l·"Public·Law·87-590·(77·Stat.·393)"·\s·"Public·Law·87-590·(77·Stat.·393)"·\c·2·⟧,·which·was·amended·by·Public·Law·95-5·86·(92·Stat.·2485)·in·1978.¶

¶

Figure 11-19

Click the Home tab of the Ribbon

Click the Show/Hide command in the Paragraph group (See Figure 11-20).

Figure 11-20

As soon as you click this the formatting marks will again be hidden from view. See that wasn't so bad! All is well with the world again.

Press Enter until the insertion point is about in the middle of the page

Click Insert Table of Authorities on the Reference tab

Another dialog box will come onto the screen as shown in Figure 11-21.

270

Figure 11-21

For this lesson we will want to insert all categories. If you had separate authorities for court cases and statutes and a bunch of others you may want to keep all of the tables separate and that would be fine. For us, we only have one authority and we can insert all types in our table.

Click OK and the table will be inserted into the document

Inserting a Table of Authorities was not any worse than any other table. The good part is Word did all of the hard work for us.

Note: you can always mark additional citations and use the Update Table command to add them to the table.

Save your work and close the document

Lesson 11 – 7   Page Numbers

Something else you will want to consider having in your document before you print it is page numbers. This should not even need an explanation. If you want to find something page numbers will help.

Open the BranTelInfo3 **document**

This document has two pages and we can add page numbers to it.

Click the Page number command in the Header and Footer Group of the Insert tab

A drop down list will slide down and you can choose where to insert the page numbers. This list is shown in Figure 11-22.

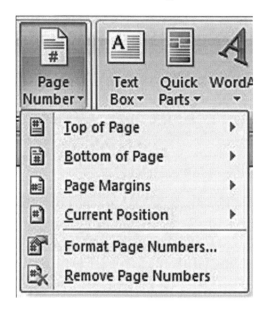

Figure 11-22

I would prefer to have my numbers at the bottom of the page instead of at the top or in the margin. The next question would likely be where at the bottom of the page would we like our page numbers?

Move your mouse over the Bottom of Page choice

Another menu will slide out to the right side showing the choices we can make. This is shown in Figure 11-23.

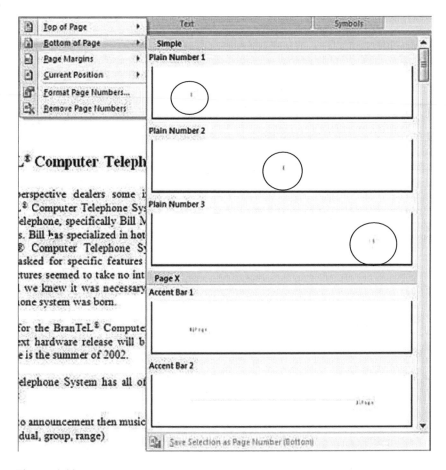

Figure 11-23

Scroll down about an inch and choose Pg. Number 2

When you click on your choice, the page numbers will immediately appear in your document.

Inserting page numbers is simple and if you have more than one page I suggest that you use it.

Save your work and close the document

Lesson 11 – 8   Inserting an Index

Almost every major document has an index in it, usually at the end, to help you find where key words are located in the document. Well we can do that just like the professional writers do. You are probably thinking that this will take several days and a lot of work to get this done. So let's find out.

Open the BranTel Info3 **document**

This document is not very long and will be easy to add a short index to the end of it. This is probably the part that you were dreading. We have to decide which words are going to be included in the index, and like the table of contents and figures we have to mark the words. Normally you will look for keyword that the user will want to find. If you were making a book of recipes you would use the name of the recipe as the keyword, not something like sugar or flour.

In the second paragraph select the words maximum size

Click Mark Entry in the Index Group of the References tab (See Figure 11-24)

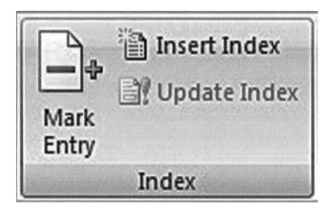

Figure 11-24

This will bring the Mark Index Entry Dialog box to the screen as shown in Figure 11-25.

Figure 11-25

The Main Entry is the selected text and there are no subentries or headings and we are going to use the current page instead of a range of pages, so all we have to do is click on Mark.

Click on the Mark button

You will notice that the dialog box stays open so you can add additional index entries. You will also notice that all of the hidden formatting marks are shown. This will happen every time you add an entry, but I still prefer to hide them while I look for the next entry.

Click the Show/Hide button on the Home tab

Select the word feature at the end of the first line of the next paragraph

You may have to click once on the document and then select the word.

Click back on the dialog box and then click Mark

If you prefer not to see the hidden formatting, you can click the Show/Hide button after every entry.

Add the following word to the index

> Support

> Certification

> Payment

> Shipping

Click the Close button on the dialog box

The words are now marked and we are ready to insert the index.

Hide the formatting marks if they are still visible

Press Ctrl and End to move to the end of the document

Press the Enter key about six times

This will ensure that there is a separation between the end of the document and the beginning of the index. Normally you will want to add a new page and insert the index at the beginning of the new page, but for this lesson this will be okay. You may have to press the Home key followed by the Backspace key to move the insertion point to the left margin.

Click Insert Index

The Insert Dialog box will come to the screen (See Figure 11-26).

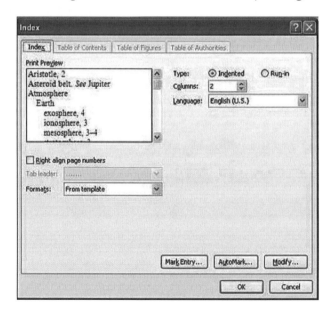

Figure 11-26

There are two changes we will want to make before we click OK. The first is we will want to right align the page numbers so they will all be even and the second is we only want to have the index in one column.

Click the check box next to right align page numbers

Use the down arrow to change the number of columns to one

Click OK

The index will be inserted into the document and should look like Figure 11-27.

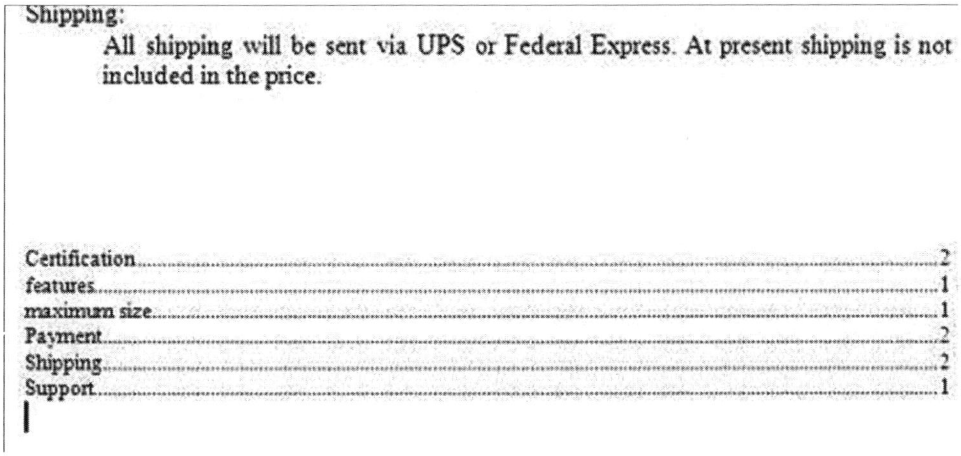

Shipping:
All shipping will be sent via UPS or Federal Express. At present shipping is not included in the price.

Certification...2
features..1
maximum size...1
Payment...2
Shipping...2
Support..1

Figure 11-27

That is all there is to adding an index to your documents. I bet it wasn't as bad as you thought.

Save the document and close it

Chapter Eleven   Review

Listed below are some of the things you might want to consider before printing your document:

1) You might want to consider adding a Cover Page. In addition to the Title you might consider adding the author's name, date, and a short summary.
2) You should decide where the page breaks fall.
3) Do you need headers and footers in your document?
4) Should you have a Table of Contents in your document?
5) Do you need a Table of Figures?
6) Have you cited authorities? If so, do you need a Table of Authorities?
7) Should you include page numbers in your document?
8) Does your document need an index?

Chapter Eleven   Quiz

1) Besides the title, what three things normally go on a cover sheet?
2) If you want to save a cover page that you have created, you must give it a unique name. **True or False**
3) In what group and tab would you find the Page Break Command?
4) A header is found where on the page?
5) On which tab is the Table of Contents Group found?
6) A footer is found where on the page?
7) When updating the Table of Contents, you can choose to only update the page numbers. **True or False**
8) A hyperlink can be inserted into a Table of Figures. **True or False**
9) If you cited several statutes (laws) in your document, what type of table would you use to show these?
10) Before you can add a word to an index, you must first mark the entry. **True or False**

Chapter Twelve   Protecting your Documents

There may be times when you don't want just anyone to read you documents. With Word you can password protect a document. If someone does not know the password, they will not be able to open the document. If you were an author and writing a new book, this would be a very good thing. This might also come in handy if you are working on a presentation for your company and you do not want another employee seeing, and taking credit, for your work.

Lesson 12 – 1   Creating a Password to Open

a Document

Create a new document

For our new document we will retype the body of the Christmas letter. It is short and easy to type.

Type the following:

Well it is that time of the year again, time for another letter from me. This is the only time of the year that you will hear from me. Well here is the latest scoop from my house. The kids are all grown and do not have time for me anymore. The economy is bad and I don't have the money to go anywhere. There is nothing new in my love life. My hair is falling out and my stomach is getting bigger. My knees are going bad and will probably need to be replaced.

Hope you have a great holiday season

From this point we can add a password to open the document, or we can add the password when we save the document. If we choose the second option, we can also add a password to modify the document.

First we will see how to add a password to open the document.

Click the Office button and choose Prepare from the menu

There are several choices available (see Figure 12-1). We are interested in encrypting our document to protect its contents.

At this point, I need to stress the importance of keeping your passwords in a safe place and choosing passwords that not just anyone can figure out.

<u>First, if you forget the password, there is no getting it back. You will **not** be able to open your document. Put your passwords in a SAFE PLACE! (And a place that you can find again)</u>

Second, if you don't want anyone to figure it out you need to use a password that is a mixture of numbers and letters. The password can be up to 255 characters.

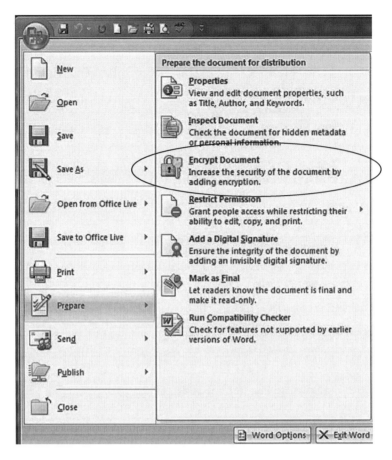

Figure 12-1

Click on the Encrypt Document

When you click on Encrypt Document, another dialog box will appear on the screen. This box is shown in Figure 12-2.

Figure 12-2

What you are going to do is enter a password that will be used to open the document. The document will not open without the password you put in the textbox.

In the textbox type 1208 **and then click OK**

A second dialog box that looks almost identical to this one will pop onto the screen. You must re-enter the same password in this box to make sure you put the first one in correctly. All passwords are case sensitive. That means that bill and Bill are completely different. You must enter the second password exactly as you entered the first password.

In the textbox type 1208 **and then click OK**

The dialog boxes will go away and everything is set. The next time you try to open the document you will be prompted to enter the password. The dialog box to confirm the password can be seen in Figure 12-3.

Figure 12-3

If you try to open the document with the wrong password, you will get an error message stating that you have entered an invalid password.

Save the document in the Word 2007 folder as Christmas Letter **and then close it**

Open the document

Instead of the document opening, you will get a dialog box asking you to enter the password.

Using the keyboard enter 1208 **and then press the Enter key**

The document will now open.

Now the only other thing to discuss is how you remove the password if you change your mind and don't want the encryption for the document.

Click the Office button and select prepare

From the slide out menu select Encrypt Document

The encrypt document dialog box will come back on the screen. This time there are four asterisks in the password text box. If you want to remove the password, you will have to delete the original password and leave the text box empty (If you wanted to change the password. You could just as easily delete the old password and enter a new one). See Figure 12-4 for the dialog box.

Figure 12-4

Press the Delete button until the password is gone, and then press OK

The document is no longer password protected and anyone can open it.

Repeat these steps and put the password 1208 back in

Save this document in and then close it

Lesson 12 – 2   Creating a Password to Edit

a Document

Having a password that will keep people from opening a document is great, but what if you want to keep someone from editing one of your documents after it is open? Word has come up with a way for us to accomplish this also.

Open the Christmas Letter **document**

You will be prompted for the password. Enter the password (1208) and click OK. The document will open and you can work on it as necessary.

Our needs have changed and we want to protect what is in the document from being changed.

From the Office Button click Save As

This will bring the Save As Dialog box to the screen.

Click the Tools Button on the bottom left of the dialog box

This is shown in Figure 12-5.

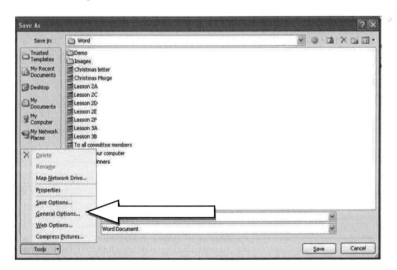

Figure 12-5

Click on General Options

A new dialog box comes to the screen, the General Options dialog box. From here we can set the password for opening the document and we can set a password for modifying the document. Figure 12-6 shows the General Options Dialog Box.

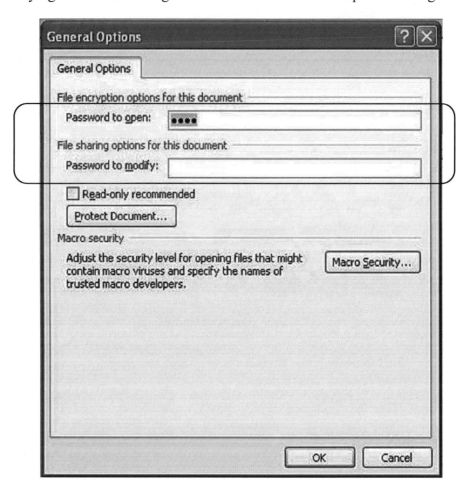

Figure 12-6

Add the password 3423 to the modify password

Now we have added a different password to modify the contents of the document. When you try to open the file, you will be prompted for the password required to open the file. If you put in the correct password, you will be prompted to enter the password to modify the contents. If you enter the correct password, the document will open and you can modify it as needed. If you do not know the password you can still open the document, but only as a read only file.

Save the document, close it, and try to reopen it

Put in the first password (1208)

When you put the first password in correctly the second password box will come to the screen, as shown in Figure 12-7.

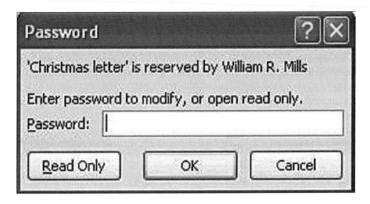

Figure 12-7

You might notice that Christmas Letter is reserved and who reserved the document. If you enter the next password (3423) the document will open. If you do not know the password and you want to see what is in the document, you will have to click the Read Only button.

Click the Read Only button

When you click this, the document will open, but no changes can be made to it.

Notice Figure 12-8 and see what a read only document looks like.

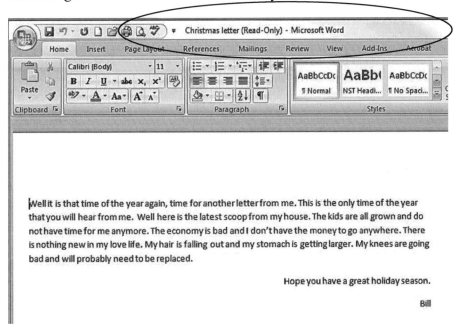

Figure 12-8

The document itself looks like any other letter. At the top, in the Title Bar, you will see that this is a read only document.

Let me explain something, you can make any changes that you want to the letter. If you try to save the letter, you will have to give it a new name, and the original letter will be just as it was when you opened it. You cannot change the original, but you can make changes to the document and then save the document under a different name.

That is how you protect your documents.

Note: The Protect Group on the review Tab of the Ribbon is not the same as encrypting a document and setting a password to open it. This uses an external company to store e-mails and documents and you have to have a Windows Live ID to use this trial service.

Chapter Twelve   Review

You can add a password to keep someone from opening a document by clicking on the Office Button and moving the mouse down to Prepare and then clicking on Encrypt Document.

If you forget or lose the password, you cannot get it back and you will not be able to open the document. Keep your passwords in a safe place and a place where you will be able to find them.

You will have to enter the password twice for security purposes.

To remove a password, the steps are similar to the steps for adding a password, only you will have to delete the existing password in the dialog box.

You can also create a password to open a document by clicking on the Tools button from the Save as Dialog Box and then choosing "General Options". This is also where you would add a password to edit or modify a document.

If you do not know the password to modify the document, you can still open the document as a read only file. Normally you would think that a read only file can never be changed, and this is true. You will not be able to save the document if changes have been made. You can, however, use the Save As command and give the document a new name.

Chapter Twelve   Quiz

1) From the Office Menu, what group or section contains the Encrypt Document command?
2) Why do you have to enter the new password twice?
3) If you enter an incorrect password while trying to open a document, you can still open the document as a read only file. **True or False**
4) Once a document is encrypted with a password, the password cannot be removed. **True or False**
5) The only way to create a password to edit or modify a document is to choose General Options under the Tools button from the Save As Dialog Box. **True or False**
6) If you enter an incorrect password to modify a document, the document will automatically open as a read only file. **True or False**
7) Neither the Save or Save Command will work on a read only document. **True or False**

Chapter Thirteen   Macros

I stated earlier that macros are used when you constantly type the same thing over and over again. With a macro you can type it once and then use it whenever you need it. This will also work if you are constantly formatting text and using the same formatting over and over. Although it is true that you could use a style for this, we are going to see how to do it with a macro.

You first have to record a macro and then run the macro.

Lesson 13 – 1   Using Macros

The first thing we have to do is record the macro. This means that we need to record every keystroke and mouse movement that is made. There is a small group on the View tab that has a button for Marcos. From this button you can record and play a macro.

Here is what we will pretend for this lesson. Don't you just love this? We can be anything we want during these lessons. It is like a being a child again.

For this lesson we are an author writing a biography about "Sir Franklin William Van Austinheimer III". This gentleman loves to see his name in print and one of his requirements is that we must use his entire name every time he is mentioned. It also must be in print at least 150 times during the manuscript. I don't know about you but typing Sir Franklin William Van Austinheimer III once was bad enough and I don't want to do it 150 times. We can create a macro to type this for us.

Create a new blank document

Click on the bottom half of the Macros button at the far end of the View Tab

This will bring a drop down list of the various options that are available. From the list we will choose the Record macro option (See Figure 13-1).

Figure 13-1

Click on Record Macro

A new dialog box will come to the screen. This is the Record Macro Dialog box and is shown in Figure 13-2.

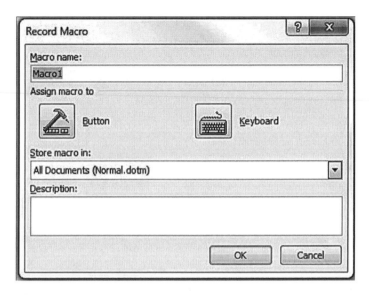

Figure 13-2

We will need to give the macro a name to identify it. We will probably want to give it a description so we will remember what it was used for in the future. We will also want to assign the macro to either the keyboard or a button so it will easier to use.

Under Macro name enter the name Biography

Under description type the following:

This is the name of Sir Franklin William Van Austinheimer III

Later we will want to use this macro (at least 150 times) in our biography story. It will be a lot easier to use if we assign it to a keyboard shortcut or button. First let's look at how we would assign the macro to the keyboard.

If you click the keyboard button, the following screen will be on your monitor (See Figure 13-3).

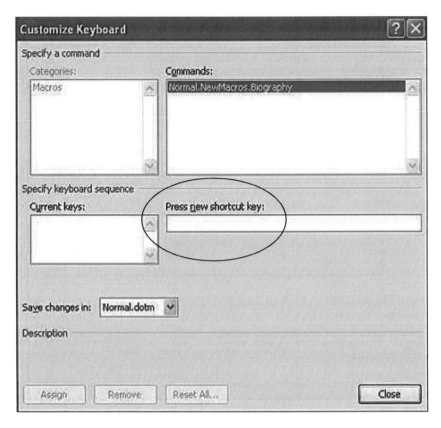

Figure 13-3

This isn't too bad, all you have to do is press the key(s) you want to use as the shortcut and then click the Assign button. If it were me, I would choose some keys that would remind me of what the shortcut was about. I might use ALT + B because it will start the Biography macro.

Okay here is the problem. You have to remember what key(s) you have assigned to the macro. You are not going to be reminded of the shortcut keys. If your memory is like mine, you probably won't want to use the shortcut keys.

IF on the other hand you decide to use the button choice, this dialog box will come onto the screen.

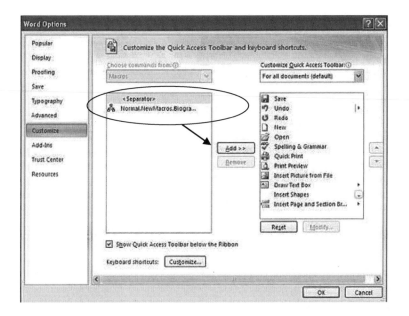

Figure 13-4

You should remember this from way back at the beginning of the book. This is the Customize section of the Word Options Dialog box. This is where you add item to the Quick Access Toolbar.

To add the item to the toolbar, you would click on Normal.NewMacros.Biography and then click the Add button. Make sure Macros is listed under the choose commands from section at the top. When you click the OK button a new symbol will appear on the Quick Access Toolbar (See Figure 13-5).

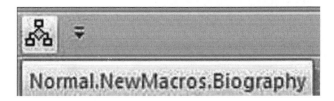

Figure 13-5

Also you will now be recording every keystroke and mouse movement that you make. This is the route we want to use with our macro.

Let's recap where we are in the lesson. We have just given our macro a name and a description and now we are going to assign it a button.

Click the button marked Button

Click on the macro Normal.NewMacro.Biography and then click the Add button

Click the OK button

The recorder will now start.

Using the keyboard type

Sir Franklin William Van Austinheimer III

Move your mouse over to and click on the bottom half of the Macro button in the Ribbon

Click Stop Recording

Now all we have to do is see if all of our hard work paid off.

You should have one line on our new document that has the name Sir Franklin William Van Austinheimer III on it as shown in Figure 13-6.

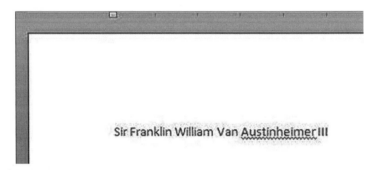

Figure 13-6

Press the Enter key to add a new line to our document

Click the Macro button on the Quick Access Toolbar

You should now have two lines in your document. Now every time you need to type Sir Franklin William Van Austinheimer III all you have to do is click the Macro button.

I said that the recorder would also record mouse movements as well as keystrokes. Let's make another macro to see if it really does record mouse movements.

Sir Franklin William Van Austinheimer III lives in the country by Manchester England. He is very particular about how Manchester England looks when it is printed. He also insists that we use the font called Monotype Corsiva and that the font is bold and slightly larger than the other words.

Add the following to the end of the second line of your document

lives in the country by Manchester England.

Now we will record a macro to change the font and size of the words Manchester England.

Click on the bottom half of the Macros button at the far end of the View Tab

Click on Record Macro

Give the macro the name Manchester

Assign the macro a button

When you are finished the macro will start recording.

If we are going to apply formatting to a particular word we must select the word before we change the formatting. We do not want the macro to include the selecting of the word, but we still have to select the word Manchester so we can change the formatting. We will need to pause the recording and then select the word Manchester and then resume the recording.

Click the bottom half of the Macro button and select Pause recording

Using the mouse select the word Manchester

Click the bottom half of the Macro button and select Resume recording

Click on the Home tab

Change the font to Monotype Corsiva

Change the font size to 14

Click the bold button

Go back to the Macro command on the View tab and select Stop Recording

Now we can see if the macro works.

Select the text England and click on the Manchester Macro

England should now match Manchester.

If you click the top half of the macro button you can select the macro that you want to run. There are also other choices that will come to the screen (See Figure 13-7).

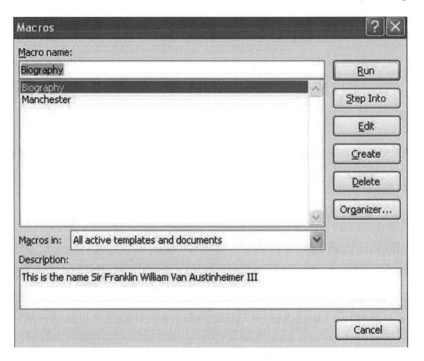

Figure 13-7

If you choose to click on edit or step into a Visual Basic screen will pop onto your computer. This is the actual programming that makes the macro work. Figure 13-8 will show the screen shot of this screen. I am showing you this so you won't be tempted to try it and see what happens. You can make the macro not work at all from here. So please don't mess with this part.

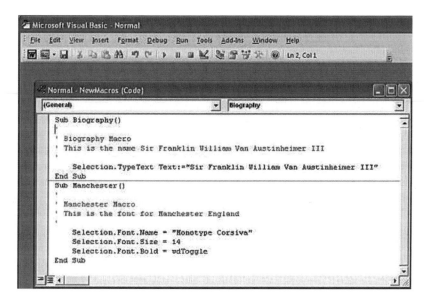

Figure 13-8

This is the actual programming that is used when you click on the macro button in the Quick Access Toolbar. The Biography part tells what text will be typed and the Manchester part give all of the formatting options. You don't have to know any of this, but I thought you might like to see it. Besides you may be so intrigued that you may actually want to study programming in Visual Basic. It really is fun.

You can also click on one of the macros and then click delete to delete the macro. Since you will never use these macros again you might want to delete both of them.

Close the document without saving it

This concludes our lessons and you should be well on your way to becoming an expert with Word 2007. Just think of how much fun it will be impressing all of your family members with your knowledge. I'll bet even the ones that said you were wasting your time and money on this book will want your expertise now.

Chapter Thirteen   Review

Macros are used when you constantly type the same thing over and over again. With a macro you can type it once and then use it whenever you need it. This will also work if you are constantly formatting text and using the same formatting over and over.

The first thing we have to do is record the macro. This means that you need to record every keystroke and mouse movement that is made.

You need to give the macro a name and a description. You will probably want to assign a button or a series of keystrokes to use the macro. Assigning a button will put a shortcut to the macro on the Quick Access Toolbar.

As soon as you click the OK button on the dialog box, every keystroke and mouse movement will be recorded as part of the macro.

To use the macro simply position the insertion point where you would like the text to be inserted, or highlight the text you want formatted, and either click the macro button or press the key(s) that were defined as the shortcut.

Chapter Thirteen Quiz

1) If you are constantly using the same formatting on text, you could record a macro to make the formatting changes for you. **True or False**

2) When you record a macro, every _____ and _____ movement is recorded.

3) All macros must have a unique name. **True or False**

4) What is the name of the dialog box where you define the shortcut key(s) for a macro?

5) You would add a button for a macro to the Quick Access Toolbar from the _____ section of the _____ _____ Dialog Box.

Manufactured By: RR Donnelley
                     Breinigsville, PA USA
                     April 2010